A
NATURAL
HISTORY
OF
COLOR

A
NATURAL
HISTORY
OF
COLOR

*The Science Behind What We See
and How We See It*

ROB DeSALLE
AND HANS BACHOR

PEGASUS BOOKS
NEW YORK LONDON

A NATURAL HISTORY OF COLOR

Pegasus Books Ltd.
148 W. 37th Street, 13th Floor
New York, NY 10018

First Pegasus Books edition July 2020

Interior design by Maria Fernandez

Library of Congress Cataloging-in-Publication Data is available.

ISBN: 978-1-64313-442-0

10 9 8 7 6 5 4 3 2 1

Printed in the United States of America
Distributed by Simon & Schuster
www.pegasusbooks.us

To RD's children, Mimi, Soso, and Bo

and all of the kids in the World.

May they enjoy the colorful world they were born into

for all of their lives.

Contents

Prologue

olor influences everything. Our universe is made up of elements that have diverse color; hence the large bodies in the universe have color. Color abounds on our planet in nonliving things. The next time you see a rainbow, or get a good look at the northern lights, or experience a beautiful sunset, remember that these are colors from nonliving parts of the Earth. Organisms evolve complex coloration patterns to warn off predators, or they use subtle colors with respect to background to hide in plain sight. Humans deal with color in a very different way than other organisms. We are perhaps the only organisms on our planet that can think about, ponder, analyze, and consciously manipulate colors.

But what exactly is color? For that matter, what is black and white? (After all, black and white are colors, right?) These questions have many answers. A painter might think about the palette of colors they can use in their work; a photographer about what colors can be captured with a camera or smartphone and how to best capture that special moment and mixture of colors. Kids might wonder at a rainbow, a sunset, or a toy shop and why certain things are the colors they are. A scientist might think about the colors coming from a star or laser beam. A marketer will use color to sell a product most efficiently. A poet will describe color with words. All of this stimulates a need to understand color and, in turn, our place in the universe.

Color is paramount to the way you experience the world around you. You can customize the mood lighting in your living room, or enjoy the programmed changing colors on some modern airplanes, or change the paint color in your children's bedroom to be more age appropriate. Color is used to influence the choices you make all the way from the grocery store to the wine shop to the art museum and the sports stadium. Color tells us what team we cheer for and what political party we prefer. Color has physical, evolutionary, social, cultural, emotional, and philosophical components. Color is almost as complex an existential concept as mortality. Color is perhaps the most complex daily neural input that humans perceive.

There is color in the objects we use, the light that reaches our eyes, the scenes we see and remember. And all this is processed by us, in our brains. That means colors are part of our way of noticing the world. Humans have picked up on color in a big way partly because we have evolved the molecular machinery to detect a thin range of light wavelengths but also because we have the ingenuity to make tools to perceive things outside our thin sliver of naturally perceived wavelengths. Color is a part of our evolutionary history written in our genes and, perhaps more importantly, in our cultures. Different colors have been instrumental in the development of the great cultures of the world. The story of how humans discovered a way to color things is a great detective story and central to the development of color's role in defining culture and in defining what humans are. Because we humans utilize color in a somewhat special way compared to the rest of the natural world, understanding what color is becomes a central question in understanding our existence.

One might argue that color began when light began. Let there be light (and colors too). But Galileo, whom we mostly remember for his astronomical exploits and near execution for heresy, pointed out that there was no hearing or vision until organisms arose that could detect sound and light. The world was both dark and soundless before the first organisms capable of light and sound detection evolved. Likewise, colors didn't exist until organisms evolved that could detect light. But this takes us back to black and white, and the idea that organisms first saw colors when they were able to discern black and white. The perception of the colors of the rainbow came later, as complex life evolved and molecular mechanisms that could split light into different wavelengths arose in nature.

The question posed above is as much a philosophical question as it is a physical, chemical, and biological one. We will establish that an understanding of color on many different levels is at the heart of learning about nature, neurobiology, individualism, and a philosophy of existence. We can pretty neatly describe the physics and biology of light and color perception, but this book will also ask questions about what it means to sense light and color.

We humans—and indeed all living organisms—are swimming in a world of information made up of small molecules, sound waves, gravity, and, most importantly for our story about color, light waves. Organisms have figured out how to use light in a wide range of ways, probably a result of the fact that there is a plethora, or a rainbow, so to speak, of different wavelengths of light hitting our planet. And a big part of the story is that there is also a bonanza of things for the light to bounce off of and be absorbed by.

All organisms use light to inform them of their surroundings, but some organisms also use it as food for energy. Plants and some bacteria have evolved mechanisms to extract energy from light. For these mechanisms a broad range of light wavelengths are gobbled up and transformed via biochemical pathways to produce energy for the plant and bacterial cells, whereas animals and fungi have figured out other ways to produce energy in their cells (though plants have this second mechanism too). Animals and fungi eat food to compensate for their other energy requirements, while plants do quite well by absorbing nutrients and through their use of light. So, light to some organisms is food and to others a source of information.

And color is one way that organisms have evolved to stretch the utility of light wave information that they are exposed to.

We humans have a stake in cutting through the information flooding us from the environment, but we have somewhat uniquely reduced the evolutionary severity of consequences of faulty or slowed processing of this information. For instance, it is absolutely imperative that a small mammal or bird in a forest process visual information nearly instantaneously and with great accuracy to ensure its survival. Today most humans simply need to know, for instance, that red at a stoplight means "stop" and green means "go" in order for them to survive an intersection. How this evolutionary give-and-take works in our species and in others is foundational to understanding color.

We can ask questions like, Does the color that you call green (and that we agree is green) "look" the same to all of us? After all, it is a pretty good bet that green is processed in somewhat different ways in our brains than in yours. Because information is processed differently in our individual brains, does it mean that your green is not our green? And what goes for colors might also go for any other information that is processed in our and the readers' brains. We might very well all live in our own little universes. Understanding color can help us understand some of these basic aspects of our existence in the universe. We can also ask a fundamental question about our planet with color. How organisms on the planet utilize photons of different wavelengths for survival is a basic theme of color perception to be explored. In fact, perceiving color is at the heart of how a lot of the organisms on our planet diversified and, indeed, also at the heart of understanding organismal diversification on our planet. Would color evolve differently in other parts of the universe? Most certainly, yes, because the physics and chemistry of the universe vary greatly, but the question then becomes, How?

It should be obvious that colors (including black and white) are a complex concept and are dependent on a variety of factors. There are four major themes in this book. The first involves an examination of what color is on the physical level. The second involves looking at the biochemical and neurobiological levels of light and light detection in organisms. The third major theme concerns color in nature and how color is used by organisms to expand the information they receive from their surroundings. How organisms on our planet diversified is partly a story of color. Adaptation and natural selection have shaped the way color is distributed on our planet and is very tightly entwined with our general impression of our planet. The final theme concerns the cultural/human context of color. But we won't be finished with color when we complete examining these four themes. Although the science we discuss will get us closer and closer to our answer to the initial question—What is color?—we still need to go back to a philosophical context, and getting closer to what color means gets us closer to what consciousness is, what existence is all about. What our existence in this universe of color means may be the most colorful story ever told.

1

The Color of the Universe

We are literally swimming in information that's all around us, which our bodies can and need to detect. We have evolved mechanisms to detect this wealth of information, from photons to sound waves to individual atoms and molecules to microbes to complex eukaryotes, all coming into contact with us. Detecting and responding to this information is essential for our survival. We and other organisms have evolved sometimes simple, sometimes elaborate, mechanisms for detecting these external stimuli. One of the most important of these mechanisms that we humans use to detect external stimuli is sight—what might be called our "overused sense." Compared to a lot of other organisms, we rely on it a great deal for survival. It will be the focus of this book, no pun intended. Color is a key ingredient, but before we get to color, we need to delve into light.

Watching the Big Bang

The grand majority of things we have at hand are what we can presently see, touch, smell, taste, and hear. However, a lot of science involves "time travel." Science looks backward at how and why these things are around us now. Science also wants to predict what is ahead of us too, as prediction is an important part of the utility of science.

Take evolution, for example. What did Darwin have at hand when he formulated his theory of survival of the fittest by natural selection? He had knowledge of the many organisms he collected and observed from his journey on the *Beagle*. He was well read and also had the knowledge of science developed before his work. Darwin wanted to understand what had happened in the past and so used these tools to come up with the most reasonable, least refutable idea about how life evolved and continues to evolve. He metaphorically travelled back in time many times to come to this conclusion. Specifically, he was able to go back in time and think about common ancestors of living things. He was also able to go back in time to visualize natural selection and how it might work to influence the evolutionary process he observed on his voyage.

Cosmologists are particularly good at time travel. They ask, What do we know about the origin of all this stuff in the universe? As with every material, beginnings can be tracked back about fourteen billion years or so to the Big Bang. As we will soon see, the only thing we know of before the Big Bang was a single point of matter so tightly compacted that it actually could not be seen with the naked eye, or any instrument.

Cosmologists study things like the expansion of the universe and have tried to tackle the biggest time travel problem of all—the Big Bang and the origin of the universe. Most of this time travel involves a unique imagination and a talent for eliminating the impossible, to detail the limits of what might have happened in the past. To address what happened before and during the Big Bang, cosmologists have developed the ultimate time machine that can not only go back in time but can also discern what happened over incredibly small intervals of time. They have concluded that the Big Bang is the origin of all the things we have around us in the universe.

The famous physicist Stephen Hawking and his colleague James Hartle thought about this in detail and did their own time traveling in order to come up with the "no-boundary proposal," to detail the state of the universe before the Big Bang. More formally known as the Hartle-Hawking proposal, it details that the universe was a singular point of mass with no initial boundaries with respect to time or space. Nice, and obtuse, right? Well actually it makes great sense, because as one time travels backward from the present, the universe compresses more and more until it shrinks to the singularity mentioned above. It shrinks to smaller than the size of a single atom, with all of the particles and mass contracted into a speck-sized clump of extreme mass and incredible heat. When the singularity state is met, time ceases to exist and definition of what happens before the singularity is, simply put, silly to think about. Everything but the singularity is closed to discussion because we have no way to define things, measure things, or even speculate about anything at that point.

Before he passed away at the age of seventy-six in 2018, Hawking was able to tackle many mind-blowing topics in cosmology. For him to conclude that the origin of time (and, for that matter, the origin of everything) is a "no-go" is quite impressive. On a TV show aired only ten days before his death, Hawking explained that going back in time to the Big Bang is a journey toward, but never reaching, nothingness. Time (and mass) shrinks more and more as it gets closer to the origin, but never makes it. As he put it in this interview, "It was always reaching closer to nothing but didn't become nothing. There was never a Big Bang that produced something from nothing. It just seemed that way from mankind's point of perspective." If one were viewing the "rewinding of the tape" of all of this time travel back to the singularity from a "safe distance," very little if anything would be visible to the human eye, and nothing would be audible or detectable by any of our senses at the end of the rewind. The Hawking-Hartle proposal, whether right or wrong, places the beginning of the tape at a singularity, an imagined situation.

Wouldn't it be cool to watch this hypothetical video run from the beginning? All it would take is sound science and a little imagination. This is exactly what cosmologists Christopher Andersen, Charlotte A. Rosenstroem, and Oleg Ruchayskiy did in their 2019 paper entitled "How Bright Was the Big Bang?" Andersen and colleagues did this by "placing a hypothetical human observer in the early Universe, and using this human visual system as a proxy for a 'light detector.'" Their thought experiment took into account the various rapid epochs that are predicted for the first second of the universe's existence. They determine two important characteristics of light in the early universe with respect to the sight our eyes accomplish: the limit of darkness and the limit of visible light. The limit of darkness is the point where complete darkness gives way to being able to see light, and the limit of light is where light becomes blindingly intense.

The first second of the video replay would be accompanied by a multitude of events, far faster than anything we can experience or measure. In a time shorter than anything we know, the singularity expanded in an event known as "inflation," when it doubled in size nearly 100 times (that is, 2100 times, or 1030 times), but at this point it was still only the size of a golf ball, and it was also unimaginably hot and energetic. As the singularity expanded, the universe cooled immensely, as energy was rapidly released, but it was still incredibly hot (109°C)—much hotter than the sun is now. At one second after the inflation, protons, electrons, and neutrons were formed. And between three and ten seconds after the Big Bang, photons appeared, spreading out from this singular point.

As expansion continued, the universe stayed so hot that photons—particles of electromagnetic radiation—moved inside the very dense soup of electrons, protons, and neutrons. The soup was so dense that the particles smashed into each other and the photons got stuck, as if in an ultra-dense fog, where light was scattered. For a million years the universe was a continuously expanding, foggy blob of particles. Finally, the universe cooled enough to form hydrogen atoms and also to allow photons to be released. Photons could move about in this transparent hydrogen soup for long distances. We still detect them as radio waves, as they are part of the cosmic microwave background (CMB) or cosmic radiation background (CRB). They are very, very weak waves, coming from all around us in space.

While the CMB is technically made up of photons, we cannot see it with our eyes. Our eyes can only detect photons in a small range of energies, or, as we say, wavelengths. Each photon has energy; if it is too high, like with X-rays, our eyes can't detect them. They pass through the body with little effect. However, when the number of photons is very high, and a high intensity exists, we can get sunburn or radiation damage. Photons with a lower energy can also not be detected by our eyes. We feel them as heat, like the warmth generated by infrared photons from the sun, or those in a microwave oven.

If we go back to Andersen and colleagues' thought experiment of the human eye viewing the video, the period of time leading up to decoupling was "blindingly bright" and full of photons. And all mainly of the wrong wavelength for the human eye to see. The human eye would only have been able to see anything once the universe was more than one million years old. As more cooling occurred over the next five million years or so, the universe became less and less bright until it reached pitch blackness to the human eye. That human eye in the thought experiment detected no light for over 150 million years, a period of the early universe called the "Dark Ages." What happened? Stars started to coalesce at this point, and enough of them formed at 150 million years for Andersen and colleagues' eye to start to detect a little light. As the universe expanded, more and more stars formed and more light was produced to get us to the current state of the universe, where there is neither too much light that would fry our retinas, nor too little light that we couldn't see.

The Top Five

Humans have seen rainbows for as long as our species has been on this planet. This includes our extinct close relatives, such as Neanderthals, who had as acute color vision as we do. All cultures use rainbows in religion and mythology. The colors in particularly vivid rainbows are among the purest visual treats ever seen by human eyes. Humans have wondered about those colors to the point of invoking supernatural explanations for them. But it wasn't until a famous experiment in the 17th century that the physical nature of the colors in rainbows was articulated in what can be called a "beautiful" experiment.

Beautiful experiments embody the essence of science. They are characterized by human cleverness and explain some fundamental phenomena in nature. According to a *New York Times* article by journalist George Johnson published in 2002, three of the top five most beautiful scientific experiments concern light and its composition. Sir Isaac Newton clocks in with the fourth most beautiful experiment focused on the nature of colors. Until Newton performed this beautiful experiment #4, scientists assumed that color was somehow a *gemisch*, or mixture of light and dark. Sir Robert Hooke, a famous naturalist of the 17th century, who liked to squabble with Newton, felt that colors were like mixing paints at a paint store. Pure white light could be mixed with varying degrees of darkness. Deep red to Hooke was white light mixed with as little darkness as possible. Deep blue, on the other end of the spectrum, was white light mixed with as much darkness as possible before the color turned black. In 1666, Newton, who never backed away from a good fight, especially with Hooke, took a simple experimental device that was popular at the time—a prism—and devised one of these beautiful experiments. Like a glass full of water in sunshine, the prism was well known to produce colors apparently by separating some special quality of the color's light. According to some scientists of the time, when light was shown through a prism, the prism itself physically altered white light in different ways to produce the many colors—red, orange, yellow, green, blue, indigo, violet, and the spread of colors in between. To these 17th-century scientists, a prism was a somewhat magical device that would "color" white light as it passed through it.

Newton had a hunch that this was a wrong way to think about colors and white light, so he used a primary prism to first get the distribution of colors normally obtained from a prism. He was then able to take the red light emanating from the prism and send it through a second prism. If the prism was coloring light, it would have an impact on the red light going through the second prism. But the color of light coming from the second prism was the same red that was isolated by the first prism. The prism was not actively coloring the light coming through it but rather was separating it into its natural components—the different colors of the spectrum. To nail down his experiment, Newton took a lens and focused the multiple colors coming from the first prism to a small point and produced white light. Not only could you take white light apart, you could also put it back together again. Newton correctly reasoned that white light was composed of all of the colors, and light was a much more complex concept than previously thought. This experiment was critical not only for the development of the physics of light but also as a guide to how science is accomplished. Many of the principles and steps of reasoning that Newton used in this experiment are still in use today.

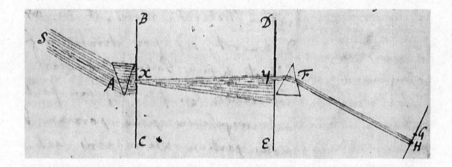

Figure 1.1. Newton's double prism experiment. Light (S) comes into the first prism (A). A small patch of separated light emanating from the prism is focused through a small aperture (X) on a divider (B-C). The light emanating from aperture X is of a single color—say, red. This pure red light then travels to the second divider (D-E) and is focused through a second aperture (Y). The light is then passed through a second prism (F). If the prism is coloring the light, as the prevailing notion went, then passing the light from apertures X and Y through prism (F) should produce a spectrum of light. It doesn't, though. The only logical explanation is that the prism is separating light, and that white light is a mixture of different hues. *Wiki Commons, public domain, https://commons. wikimedia.org/wiki/File:NewtonDualPrismExperiment.jpg.*

While Newton uncovered the complexity of white light, he also developed some ideas about what light was composed of. He felt strongly that light was particulate; it was his gut feeling, though, and not backed up terribly well by data. Newton was right about a lot of things, he usually provided data or strong theory to shore up his conclusions. Not so with light as a particle, though, and since no one is right all the time, science takes over. Here is where the fifth most beautiful science experiment comes in. There was another side to determining what light is (there is always "another" side in science before experimentation occurs) at Newton's time. Many scientists felt that light behaved like a wave. And there couldn't be two more different ideas about the makeup of something.

For a lot of us, tossing stones into a still pond is a pleasing endeavor, mostly because of the waves produced. For a very long time, children and adults have enjoyed this pastime, because, like rainbows, the effect probably enthralled them. Sound was also a big-ticket subject in science around the 17th and 18th centuries. Both sound and water waves were studied and characterized in this period of time, which was important for establishing methods of science. The scientists who studied these things used observation and simple mathematical modelling of what they saw. What they discovered about waves is seminal to understanding light even though, as we will see, the explanation isn't so simple.

Waves have very specific patterns of behavior, and the characteristics of waves were worked out well before the 19th century. If we look at a wave on the ocean, we can see that it has high points and low points. The former is called a "crest" and the latter is called a "trough." The distance between the trough and the crest is called the "wave height." Half of the wave height is called the "amplitude" of the wave. This terminology makes waves seem like they are higher than they really are. A wave with a height of ten feet sounds pretty big, right? But if you are watching this wave in the ocean, you will note that it rises no more than five feet above the surface of a smooth ocean. The trick is that the wave also sinks five feet below the surface of a smooth ocean. The distance between crests is called the "wavelength" and is represented by the Greek letter lambda (λ).

If you are watching those waves on the ocean closely, you will eventually see that some are coming in at angles to others and they crash into each other. When they do hit each other, the simple wave patterns are disrupted and a phenomenon called "interference" occurs. If two waves collide with each other at their crests, the amplitude of the wave is bigger than either of the two waves by themselves. If a trough of one wave hits a crest of another, the result is smoother water than either of the two waves produce on their own. What is happening here? It turns out that the waves are in effect adding up their individual effects, and waves that represent the sum of the two individual waves are produced by the interference.

In 1803, Thomas Young, a British physician-polymath, devised the fifth most beautiful experiment. He would force light through a pinhole and manipulate it with various objects, like mirrors or cards. Scientists create novel devices all the time, and the verb "play" is not too far off from what they do with their invented devices. Sitting in a dark room, Young shone light through the pinhole. He took a small card (as he describes, about one thirtieth of an inch thick) and used it to bisect the thin beam of light along its path of projection. He then placed a screen at the end of this "device," to visualize the effect of splitting the white light. Because the two beams of light separated by the card were from the same source, Young reasoned that if light was a particle, then once the two streams of light were created, they would produce two separate point streams of light. If light was a wave, then the two secondary beams of light would interfere with each other like waves. Indeed, the result of this beautiful experiment was a pattern showing alternating light and dark bands, behaving just like waves would when interference occurs. Where the two beams overlapped their crests, they reinforced each other and made lighter bands. Where the two beams collided one at its crest and the other at its trough, the light from one beam was cancelled out by the other, producing a dark band. Over the years, scientists learned to use a card with two holes instead of a one-thirtieth-inch card to split the single light beam. These experiments are called "double slit experiments," and they are how more recent experiments in quantum theory were conducted to establish wave or particle behavior of physical phenomena.

While the work of Christiaan Huygens in the 1600s is not listed in the top ten experiments because his contribution was mathematical, it is a major contribution to how light behaves. In his *Treatise on Light*, he developed a mathematical theory that described light as wavelike. Young's beautiful double-path experiment established that light might very well behave like a wave, but it did not eliminate the possibility that it was also particle like. The only conclusion to come to is that neither Newton nor Young was right. Or another way to say this is that neither Newton nor Young was wrong. At the beginning of the 20th century, Albert Einstein and Max Planck demonstrated that light was made of packets of matter called "photons," taking us back to a more particle-based definition of light. But the double slit experiments kept coming, and these led scientists to hold onto waves as an explanation. In 1929, Arthur Eddington, a British physicist and astronomer, with a little quantum theory suggested that light is made of "wavicles"—stuff with the characteristics of both waves and particles, with complementary characteristics at the same time. This helps a little, but it is not entirely correct.

It is kind of funny that the most beautiful experiment of all time was a thought experiment. With quantum mechanics in hand, one could think their way through the problem in the following way. Imagine setting up a thin beam of electrons. By using the equations of quantum mechanics, an original stream of electrons could be split by a double slit placed in its path. Quantum mechanics equations give only one result of these split electron streams—they will interfere with each other. The electrons (clearly particles in this experiment) would leave the same interference pattern as light split by a card or forced through a double slit. No other way to explain it in the context of quantum mechanics, and so particles could act like waves. One thing that we all should learn about physics, and astrophysics in particular, is that oftentimes things can be right and wrong at the same time. They are, as some physicists say, "what they are."

It is pretty amazing that it took about forty years to actually physically perform this most beautiful of all beautiful experiments. It took this long because researchers simply did not have the right tools to perform it. Why? A beam of electrons is an incredibly tiny thing in cross section, and finding something to split this incredibly tiny beam took ingenuity. And it didn't happen until 1961, when Claus Jönsson devised a way to split an electron beam with primitive nanotechnology. Taking this most beautiful of experiments a step further, Italian researchers decided to split a single electron and see what would happen. Believe it or not, quantum mechanics predicts that when a single electron is split, it will interfere with itself. And lo and behold, in 1974, Giorgio Merli, Gian Franco Missiroli, and Giulio Pozzi demonstrated this very result.

If electrons behave this way, then photons should also. But there is a distinct difference between electrons and photons. Electrons have mass and charge and behave entirely like particles. Photons are thought to have no mass, are not charged, and have this strange property of wave behavior. All of these differences make electrons suited for certain activities and photons for others. And when it comes to life as we know it on Earth, these roles are incredibly important for sensing the outside world. As we will soon see, electrons have become the currency of neurological processes and photons the currency of light as it impacts most organisms on this planet, including vision in animals. These processes all shook out this way as a result of evolution. It's not that evolution took photons and electrons and changed them into something we now observe as part of organismal diversity; rather, photons and electrons were always there, and evolution figured out a way to use them resulting in how organismal diversity on our planet coped with them. How this happened is the subject of the rest of this chapter.

Blackbody and Background

When it comes to the universe, the things out there that we know something about make up less than 5 percent of the total stuff out there. Our universe is made mostly of dark energy and dark matter, appropriately named "dark" because we know very little about them. At least we know about the darkness, and that means something is there. Luckily for our story about color, this 95 percent of the universe is, so far, irrelevant. What that leftover 5 percent is made of, though, is pretty important.

You might think that our sun is the source of most of the photons that we are exposed to on our planet. After all, it is the brightest thing in the sky, and when the sun slips over the horizon at nighttime, we have only the light of stars (and cyclic reflected light from our moon). But we would be wrong in assuming our sun produces most of the light we are bombarded with. Instead, the CMB is the most abundant source of "light" on our planet. We don't really see it with our eyes, as it is—as its name implies—microwave radiation. If we were able to visualize microwave radiation as well as visible light, our vision would include a very hazy background light that would be the CMB. It has been estimated that only one in every 1,029 photons out there is not CMB. In other words, a very, very small proportion of photons are not microwaves.

Scientists oftentimes need to create idealized worlds and objects in order to explain things. One idealized object in physics relevant to our discussion is a blackbody, or a physical object that absorbs all of the light hitting it regardless of wavelength, frequency, or angle of incidence. Now, think of a true blackbody. It reflects no light whatsoever. But it will radiate in a very specific way, giving off a very specific spectrum of wavelengths of electromagnetic radiation—specific because the blackbody's temperature will dictate the wavelength of light emitted by the blackbody. Even though our sun is called a "radiator," it really isn't. And here is where physics gets weird with words again. We start with the fact that a perfect absorber of light will also be a perfect emitter of light. Sounds like we are trying to sell you a bridge in Brooklyn, right? But it actually makes a lot of sense. Think of how light gets absorbed by a body like the sun. A body of matter is made of atoms, themselves made up of smaller particles like protons, neutrons, and electrons. These particles have electrical charges, which is a simple property that dictates how the particle will move in an electromagnetic field. When photons crash into atoms, some of them pass through the atoms, and in doing so push the particles around. This movement raises the energy of the atom; since the total energy of a system like this needs to stay the same no matter what happens, energy has to be released to balance out the energy dynamic. The release of energy comes in the form of photons with different wavelengths.

The sun itself has no solid surface, and when light, or any kind of radiation, smashes into it, the radiation is both scattered and absorbed, making the sun appear very much like a blackbody. Close to a blackbody, but not like a perfect blackbody radiator, because as the sun absorbs radiation, and light specifically, atoms of the sun block some of the radiation, increasing the total energy of the sun. To account for this increase in energy, the sun emits radiation, or light, at various wavelengths. Radiation that hits the sun then generates photons of various wavelengths that make their way from the sun's surface and through and beyond our solar system. The important thing is that a great deal of this radiation hits our planet and the entirety of it is blackbody radiation of specific wavelengths determined by the temperature of our sun. As we will see later, it is the temperature of a radiating body, like our sun, that determines the kind of light that is emitted from the radiator.

Two very recent calculations become relevant to our understanding of light in the universe and how much of it reaches our eyes. The first concerns a calculation made in 2018 of the number of photons that have been produced so far in the universe. The details of how this was done involve blazars (galaxies with super-massive black holes), epochs (periods of time in the 13.7 billion years of evolution of the universe), the cosmic fog (that plasma we discussed earlier in this chapter); and NASA's Fermi Gamma-ray Space Telescope. The number turns out to be $4(10^{84})$. Oh, what the heck, let's just write it out—4,000,000,000,000,000,000,000,000,000,000,00 0,000,000,000,000,000,000,000,000,000,000,000,000,000,000,00 0,000 photons.

The second calculation was made in 2013, by scientists at the Max Planck Institute who calculated the minimum lifetime of a photon. Their estimate was an average of 10^{18} years, or perhaps more easily written as a billion billion years. The age of the universe calculated from the Big Bang is 13.7 billion years, so most photons in our universe are nowhere near "dying." Because the universe cooled as it expanded (it is now only 2.7° C above absolute zero), the amount of light reaching us as CMB was also diluted. Of those non-CMB photons reaching our eyes, the remainder do indeed come from our sun in the form of blackbody radiation. Needless to say, the number of photons in the universe is an incredibly large number ($4[10^{84}]$). Even though a large number of photons reach our eyes each day, the light we see is pretty dim compared to what it could be. But because photons are pretty resilient, it means that even with the vastness of the universe, many of these photons reach our eyes daily. And it is those photons that we now turn to for color.

Wavelengths

Electromagnetic radiation is made of photons, which have wavelengths, and these wavelengths determine the physical characteristics of the photon. There is no theoretical upper limit for electromagnetic radiation wavelength, and the largest one in human use so far involves ultra-low-frequency (ULF) radio waves. This kind of radiation is used by the military to communicate with submersible craft. The exact details are more than likely classified, but this radiation probably exists in the 100 km to 1,000 km range. Before we go any further, we should standardize the wavelengths that we will discuss, but if you are adverse to exponential notation and powers of ten, skip to the next paragraph. The standard unit of measurement for wavelengths is nanometers, because visible light has wavelengths between 400 and 700 nm (1m equals 1,000,000,000 nm, or 10^9 nm). So, these ULF waves have wavelengths on the order of 10^{13} nanometers. To give you a better idea of the wavelengths of radio waves, AM radio relies on photons with wavelengths of 10 meters, or 10^{10} nm, and FM radio radiates photons at wavelengths of 1 meter, or 10^9 nm. These are photons with huge wavelengths compared to visible light.

The microwave ovens in our kitchens use photons generated at fairly long wavelengths but quite a bit shorter than radio waves. Your microwave is generally made of a magnetron, a waveguide, and a cooking chamber. The magnetron is a tube with a magnet and an electric current that generates photons of wavelengths on the order of 10^9 nm. The radiation from the magnetron acts to create molecular movement in the object it is bombarding. This movement generates heat as molecules in the object in the food chamber slam into each other. The movie franchise Predator has a great example of the next category of photons we need to examine. In the original film, Arnold Schwarzenegger, playing an Army Special Forces combat soldier, encounters a nasty alien hunter in the jungles of Central America. It stalks and kills everyone in Arnold's troop. Arnold escapes in typical Arnold fashion because he realizes that the alien is using the warm-bloodedness of his prey to visualize the members of the troop. By smearing cold, wet mud all over his body, Arnold tricks the visualization device that the Predator uses to locate prey, and he is able to extract vengeance on this particularly ugly alien. How did the mud confuse the Predator? Body heat gives off radiation at specific wavelengths. These are photons that have wavelengths on the order of 10–5 meters, or about 1,000 to 10,000 nm. The mud lowered Arnold's surface temperature, which then allowed him to blend in with his surroundings, which were mostly plants with cool temperatures.

Visible light, over which we see things with our eyes, resides in the range of 400 to 700 nm. This thin slice of about 300 nm in range accounts for all of the colorful world that most of us can perceive. Even shorter wavelengths exist than the visible range; these are ultraviolet radiation (UV), X-rays, and gamma rays. UV radiation wavelength ranges from just outside the short end of the visible range to about 10-8 meters, or 100 nm to 1 nm. X-rays are on average 10-10 meters, or 0.01 nm. We all know the typical use of X-rays as a way to visualize things in our bodies that reside (or were placed) under our skin. An X-ray can do this because these photons of 0.01 nm wavelength can penetrate the skin but get absorbed by bones and other things under our skin. Gamma rays are made up of photons on the order of 10-14 meters, or an incredibly small fraction of a nanometer (0.000001 nm).

This journey across the range of electromagnetic radiation might not mean much to those of you who are exponent challenged, so let's compare the wavelengths of these various kinds of radiation with everyday items. ULFs have wavelengths the size of the distance from Detroit to New York City. FM radio waves are made of photons of wavelength about the same length as a yardstick. Microwaves have wavelengths about the diameter of a US dime, and infrared wavelengths are the same length as a bacterium in your gut. Visible light correlates to wavelengths the same size as the average width and length of a pit on a compact disc (CD), which has something to do with why CDs reflect light in a rainbow of colors. UV light has wavelengths on the same order of size as the thickness of the cell wall of that bacterium living in your gut. X-rays have wavelengths the size of the width of a sugar molecule, and gamma ray wavelengths are so short there are very few recognizable analogues for the layperson. We can say, though, that gamma ray wavelengths are about four orders of magnitude shorter than the diameter of a hydrogen atom.

Since this book is about color, after all, let's return to that 400 to 700 nm sliver in the range of electromagnetic radiation that is visible to our eyes. At the longer end of this tiny sliver, just outside 700 nm radiation, is light that is called "near infrared." Just shorter than near infrared are wavelengths from visible red light. Various shades of red are produced by photons of wavelengths from 600 to 700 nm. At the short end of this sliver is the light that is purple/blue to our eyes. This tiny sliver of the visible spectrum ranges from 400 to 500 nm. In between, with wavelengths between 500 and 600 nm is the light that appears as orange, yellow, and green colors to us. It is amazing that our visual system and those of most other organisms on the planet rely on this tiny sliver of the entire spectrum of electromagnetic radiation. Of all the wavelengths ranging over twenty-four orders of magnitude that organisms on this planet are exposed to, by far visible light wavelengths are the most popular kind of light that organisms on our planet have exploited for the extraction of information. Visible light that produces color doesn't even range over a single order of magnitude. Does this mean that color vision on our planet is something special requiring metaphysical explanation? As we will see, the answer is no, because the process of evolution can provide a plausible explanation for why this narrow range is used by organisms on our planet. The breadth of radiation that exists out there is amazing, and settling on this tiny sliver to interpret color might seem astounding at first. But evolution has led to the use of this tiny range of wavelengths in an easily interpreted pattern of exploitation.

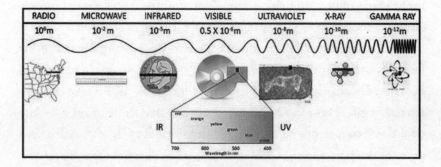

Figure 1.2. Distribution of electromagnetic radiation. The category of wavelength is at the top in meters (m). Visible light range is shown at the bottom in nanometers (nm). See text for discussion. *Drawing by Rob DeSalle.*

When Did Color First Appear in the Universe?

This question at first might seem like a chicken and egg problem. Many chicken and egg problems are not solvable. Take the origin of life on our planet. It is undeniable that all life on this planet arose from a single common ancestor. It's undeniable because of the similarities of certain aspects of the biology of all living organisms on Earth, which are based on the interconnectivity of the genomes of organisms on this planet. As of 2018 the genomes of tens of thousands of different species have been generated. It turns out that there is a core of genes that are required for cellular life on this planet, and this core of genes (between 100 and 200 genes) is found in all of the cells of all of the organisms on this planet. The easiest explanation for this somewhat stunning fact is that these genes existed in a common ancestor. It is hard to deny that all life on the planet is related, but where did this common ancestor come from?

Studies focusing on the origin of life are not designed to answer the question definitively. They are designed to eliminate possible explanations and to give science an idea of the range of possibilities that could be involved in the origin of life. No single origin of life study or theory has definitively answered the question. One theory proposes the origin of life on our planet to be a completely chemical process wherein the basic molecules of life form from smaller components organically or with some environmental stimulus. Others suggest that the origin of life on our planet is of extraterrestrial origin. This idea, called "panspermia," doesn't necessarily depend on little green men but rather on the simple introduction of a single-celled ancestor through cometary delivery. Panspermia does not solve the chicken and egg problem, and the chemical explanation only lives on because it resists rejection. Some problems in science are either incredibly difficult to answer or are unanswerable, but they are so darn interesting that scientists just can't resist trying to tackle them. Understanding what happened before the Big Bang, which we discussed earlier, is another example of one of these irresistible problems.

It would be imprudent to deride the claim that somewhere in the universe entities have existed that evolved the capacity to detect photons of different wavelengths. If this has happened, then color in our universe would have first appeared at that point in time and in that place. Since we have no access to data to test this hypothesis, our hands are tied in further exploring the origin of color in the universe. The origin of color escapes being unanswerable because we can theoretically and experimentally pinpoint the origin of light and color detection by organisms on this planet. In other words, the problem is a simple evolutionary one. However, we can only pinpoint this evolutionary event for our planet so far. In other words, color first appeared in the universe when organisms evolved to discriminate between photons of different wavelengths. In terms of how we look at color, this reliance on detecting wavelength becomes our working definition for color on our planet. As we will soon see, photons can range over a wide swath of wavelengths, and what appears to be colorless—even invisible—to our vertebrate eyes can be a rich source of wavelength information for other organisms.

Natural Selection

Understanding evolution on this planet lies at the heart of understanding what color is. There is every reason to think that organismal evolution also occurs in other parts of the universe, but all we can really address at this point in time is the process here on our planet. The best assumption we have is that gravity is a pervasive and uniform force over most of the universe. This guides the discipline of physics in explaining how the universe works. There is every reason to think that evolution too is a pervasive force in the universe. Charles Darwin and Alfred Wallace, two 19th-century naturalists, provided science with the rationale for why evolution was pervasive in nature. Naturalists at that time understood that organismal life evolved, but they had no idea how it might work. Darwin and Wallace provided the mechanism for how evolution works.

One of Darwin's seminal contributions to science was providing an explanation for what is needed for organismal evolution. Darwin made the amazing observation that organisms themselves don't evolve, they are just born, they reproduce (if they are lucky), and they die. Lamarck, a famous 18th-century naturalist who lived a bit before Darwin, made the mistake of thinking that organisms themselves evolved. He presented the notion that acquired characteristics were involved in how evolution proceeded.

Darwin changed all of this by demonstrating that populations evolve and not individuals. Once he realized that populations evolved, he was able to put in place several requirements for how evolution proceeded. Darwin pointed to the importance of variation in nature as a major requirement for evolution to proceed. In fact, Darwin was infatuated with variation, as his studies of pigeons and dogs clearly show. He described variants of pigeons in gory detail in *On the Origin of Species* as an example of the concept of variation. He also recognized that reproductive success of some individuals over others was an important cog in the process of evolution. If that wasn't enough, Darwin also recognized that organisms produce more offspring than are needed for the next generation. Because some individuals are more successful than others at reproducing, Darwin also pointed out that resources for survival and reproduction must be limited for evolution to work. Finally, and most importantly, the traits that vary in populations need to be inherited from generation to generation. With all of these important processes in nature recognized by Darwin and described as one long argument in *On the Origin of Species*, he then named the process that would take over if these requirements were met in a population. He called it "natural selection."

While Darwin's outline of the general requirements and how they interact was absolutely spot on, there were some holes in his representation of evolution. These shortcomings were not necessarily his fault, and by no means were they deal breakers for Darwin's suggestion of natural selection as a pervasive force in nature. As the philosopher and historian of science Thomas Kuhn so adeptly pointed out, science proceeds episodically, with periods of stasis, when normal accumulation of scientific evidence proceeds, interrupted by sharp periods of revolution, when basic paradigms of research shift. Indeed, Darwin's formulation of natural selection was revolutionary, as the last 150 years of research in natural history and evolutionary biology demonstrate. Researchers have used natural selection as a fulcrum for massive amounts of what Kuhn would call the normal science of accumulation. The information accumulated over the past 150 years of evolutionary research has filled in many of Darwin's shortcomings, especially the lack of a sound mechanism for heredity.

Gregor Mendel, the "monk in the garden," was a contemporary of Darwin's. Mendel worked tirelessly at understanding the transmission of traits, and hence heredity, through the many crosses of pea plants he did in the Moravian monastery where he spent a great deal of his life. Scientists back then communicated much like scientists do today—in principle, though not in practice. Unlike today, when two scientists can quickly communicate with each other via email and send PDFs of their work, scientific communication back in Darwin's time was cumbersome. Actually, this tradition of scientists ordering reprints (copies of the papers they publish) from journals and sending them to colleagues survived until the 1990s, when the internet made it very easy to simply send an electronic copy. More than likely, there are still some old-school scientists out there who still package their reprints, address them, and send them via snail mail to their colleagues, but it has been a long time since either of the authors of this book have gotten one of those manila envelopes in the mail.

Fortunately, Darwin kept all of his scientific literature and even marked the margins of the papers with notes. It is controversial as to whether or not Mendel sent Darwin a copy of his paper published in 1866 entitled "Versuche Uber Pflanzen-Hybride" (Experiments in Plant Hybridization) describing the work. It was published in an obscure natural history journal from Brunn, Czechoslovakia (now Brno, Czech Republic). Robin Marantz Henig, in her book *A Monk and Two Peas*, suggests the paper made its way to Darwin's library. She based this conclusion on the testimony of a former director of Mendel's library, who claims that a reprint was sent to Darwin. But perhaps it was not received because there is no record of the reprint existing in the Darwin library.

Other historians are resistant to this suggestion. David Galton gives the rundown on the location of the forty reprints that Mendel ordered, and there is no evidence that a reprint made its way to Down House, Darwin's residence near London. Galton points out that Darwin did own a book about plant hybrids, by German botanist Hermann Hoffman, where Mendel's results were reported. Nelio Bizzo and Charbel N. El-Hani point out he did mark the margins of that book, indicating he read the book and was not dozing when doing so. Even though the notes are clear, Bizzo and El-Hani suggest that either Darwin did not understand what Mendel had done, or it was deemed by Darwin as peripheral to his ideas about natural selection. On the other hand, Mendel ordered a copy of Darwin's *On the Origin of Species* and marked it up in the margins. It is clear from Mendel's marginalia that he did not make the connection between his work and Darwin's natural selection. Mendel's work slipped into obscurity soon after it was published until the turn of the 20th century. In fact, the botanists who resurrected Mendel's laws are said to have "rediscovered" them. But once rediscovered, Mendel's laws, when applied to evolution, open up a broad swath of work and were involved in the creation of the subdiscipline of population genetics. Genetics became the foundation for much of evolutionary thought in the 20th century, and indeed genetics is intricately involved in how we will approach the evolution of color.

With heredity in hand, other aspects of the requirements needed for natural selection can be explained. Specifically, genetics can explain how variation is generated. In addition to population genetics, established in the 1930s, heredity was cemented to evolutionary processes in the new, or Modern Synthesis, of evolutionary thinking developed in the 1940s. Heredity is real, and its addition to evolutionary theory created a sound formalization for studying organismal change. An understanding of how variation is generated is another product of the study of heredity. The grand majority of organisms on this planet have only a single copy of their genome in their cells (protists, bacteria, and archaea). However, most organisms that can detect color have two copies of their genome in each cell (eukaryotes)— one from their mother and one from their father. Mendel's laws of inheritance were formulated for these organisms with two copies. The microbial organisms on this planet that have only a single copy of the genome in their cells reproduce by fission, a process that only needs an understanding of clonal inheritance to be understood. Variation in these microbes is produced when fission occurs and the new genome of the daughter cell is synthesized. The synthesis of the new genome is implemented by enzymes in the cells of these single-celled organisms and the faithful replication of genes by these enzymes is not entirely perfect. We will discuss the nuances of the genetic material in the next chapter, but suffice it to say that mistakes can and are made in the new genomes, albeit at a very low rate. Think about it: if there was a high rate of mistakes, then a large number of changes would occur, and this might harm the viability of the daughter organism. A low rate of change will indeed produce some changes that are very deleterious to the daughter organism, but the low rate decreases the frequency of these deleterious changes relative to a higher rate. The outcome is a daughter cell that has an alteration in the makeup of its genome. Such changes in the genome are called "mutations."

Eukaryotes, even single-celled ones, need a special mechanism to implement the coming together of the genomic contribution from mom and dad. The mechanisms that nature settled on for eukaryotes are various, simple, and elegant. Suffice it to say that not all eukaryotes reproduce the same way. But at the heart of their reproduction is the same theme. Mom produces cells (oocytes), which come together with cells produced by dad (sperm), to produce a new organism (zygote). The oocytes and sperm are collectively known as "gametes," which contain only a single copy of mom's and dad's genomes, respectively. Something interesting can happen in the generation of gametes. You probably guessed it. As gametes are formed, the enzymes that copy the genome make mistakes and produce mutations. Again, not a high rate but high enough to produce variation in a population of gametes. Once the variation is generated by the way genes and genomes are inherited, natural selection can then take over. One thing to remember about natural selection is that it is environment dependent. Natural selection will act differently in, say, a cold environment than it will in a hot environment. We will delve into this aspect of natural selection when we discuss coloration variation in natural populations of organisms. When a particular population stabilizes in an environment by increasing a particular kind of a change, the population is said to have adapted to its environment.

Perfection?

There are shortcomings to the application of natural selection in the study of organismal diversity though. As pointed out above, the 20th century was a period of normal science in evolutionary biology, when researchers hammered the paradigm of natural selection into natural history. Actually, they over-pounded it, and this prompted two famous biologists at Harvard University to set the record straight. Richard Lewontin and Stephen J. Gould pointed out that evolutionary biologists had fallen into a trap they called the "adaptationist program." Researchers had, according to Lewontin and Gould, relied too much on natural selection, and hence adaptation, as an explanation for organismal diversity. Everything appeared to evolutionary biologists at the time as adaptations, and a lot of misguided ideas about what natural selection could do were promulgated. What this caused was a lot of "just so" stories and a lot of evolutionary storytelling. Many misconceptions of adaptation arose, among them the idea that adaptation produced perfection. Lewontin and Gould's purpose was to clarify the role of natural selection in nature and to define more precisely the way adaptation worked. They were mostly concerned with the overuse of adaptation in human evolution. To them, the prevailing paradigm of adaptation explaining a lot of human biology and behavior in particular was misguided. These caveats to an adaptationist approach to nature are compromises, spandrels, kluges, and drift.

Compromises: They specifically pointed to four areas of research on adaptation that were problematic to them in the context of biological perfection, or even optimization. The first is that some traits that we see in nature are compromises. Basically, they were pointing to the difficulty in studying complex traits like morphology and behavior and making simple statements about adaptation. A good example of adaptational compromise is seen in many organisms we are familiar with. A seal's flipper, for instance, seems to be nicely adapted to swimming in the ocean, but it's rather cumbersome when seals do one of the most important things in their lives. Seal flippers are a biomechanical compromise to allow seals to move around on land during reproduction and to swim swiftly underwater. As many of us well know, our knees are fragile, poorly designed structures. We know this because we have either torn knee ligaments, busted kneecaps, or live in continual pain, not only in the knees but also in the back as a result of mis- and overuse of this joint. A properly designed human knee would look more like the ankle joints of a bird's leg, according to S. Jay Olshansky, Bruce A. Carnes, and Robert N. Butler, who first suggested the design, and Alice Roberts, who interpreted it into three dimensions with her "perfect" woman 2.0. The perfect knee would actually be one where the joint is reversed. In fact, Olshansky and colleagues note that to survive to 100 years, our bodies should indeed have this birdlike leg among other physical alterations like larger ears, a curved neck, and shorter limbs. A bird's knee is usually pretty far up its leg, hidden by feathers. The ankle of a bird is apparently more robust and allows faster running than the human knee. The claim that the human knee is perfect is quite erroneous, then, as it is really a compromise between standing upright and wear and tear on our knee joint.

Spandrels: Next time you are looking in the mirror, take a close look at your philtrum. It's that area of your face between the bottom of your nose and your upper lip. If adaptation has a role in structures, you might be tempted to come up with a role for this part of your face. Maybe the philtrum enhances the moving of odors hitting your face into the nose. Since odors are an important signaling system for organisms, this function would be adaptive for an organism with a philtrum. On the other hand, maybe the philtrum exists because it allows for some room between the nose, which occasionally drips harmful bacteria, and the mouth, so that humans can easily wipe away the harmful bacteria without swallowing them. Or maybe both of those scenarios are evolutionary BS, because perhaps the philtrum is there because this is just the way the human infant face develops. Lewontin and Gould called such features like the philtrum "spandrels," in reference to the architectural spandrels of San Marco (St. Mark's Cathedral in Venice). The spandrels in this wonderful piece of architecture circumscribe beautifully drawn and painted biblical artwork that appears to be perfectly adapted to the spaces circumscribed by the spandrels on the cathedral's ceiling. The mere beauty of the artwork and its placement tempts one to think the spandrels were created to hold the artwork. But this would be an erroneous assumption, because the spandrels were concocted as an architectural device to hold the ceiling up.

Kluges: Some traits we see in organisms simply cannot evolve because their starting points are constrained by the variation that is present in the populations of those organisms. If you're trying to make a better mousetrap and need a particular piece, like a stronger spring or a preferred metal base instead of a wooden one, you simply go to the hardware store or get online to order the exact piece needed. If those pieces aren't available, then sometimes you end up with the convoluted Rube Goldberg apparatus in the popular child game Mousetrap. Populations of organisms do not conjure up variation upon request. Adaptation acts with the variation that exists in natural populations. This, by the way, is why evolution oftentimes produces convoluted, messy solutions to natural challenges that are far from optimal . . . but work. Gary Marcus has called such structures "kluges." They have come to be known as something not built according to design but assembled from whatever is available to make things work. Marcus calls our brains kluges because of their messiness with respect to behavior, but after all, they do work.

Drift: And finally, some traits arise in populations because of darn simple luck (good or bad). One of the major things that Darwin missed in his formulation of evolution was that a phenomenon known as "sampling error," or genetic drift, can occur. This concept is best explained by a coin flip example. If we wanted to bet you that we could flip a coin and land on heads two times in a row, you could conceivably take that bet, because you would lose only a quarter of the time. If we happened to hit a lucky streak, we could win a lot of money from you though, but such a streak would be rare. On the other hand, if we wanted to bet you that we could flip one hundred heads in a row, that is a bet you should take every time. The probability of flipping one hundred heads straight is minuscule (1/2,100) and so an excellent bet for you. The larger the sample, the smaller the sampling bias. The smaller the sample, the higher the probability that a strange result will happen. Likewise with populations. If a large population incurs a mutation and then is allowed to evolve, the gene with the mutation can only increase if natural selection favors the mutation. On the other hand, if a mutation arises in a very small population, it can increase in frequency due to sampling error even if it is deleterious. On the other side of the coin, if the population is very small and the mutation is advantageous, it can be eliminated from the population due to sampling error. Strange things can happen with small populations, and indeed we see some human populations where high frequencies of deleterious genes exist as a result of sampling error or, more accurately in this case, because of genetic drift. The remnants of small population size effects are seen in current-day Ashkenazi Jewish populations, where such disorders as Tay-Sachs disease and Canavan disease are found in unusually high frequencies as a result of a history of past population constrictions within this ethnic group.

Color Vision in Karnet of the Betelgeuse System

The reason we talk in so much detail about evolution in the context of color is to help explain how light and color detection have molded organismal diversity on our planet. As we pointed out above, the environment is the arbiter of how natural selection will work. A population of organisms will respond to an environment with constant freezing temperatures differently than it would respond to an environment with constant temperatures over 120° F.

While the naive amateur stargazer probably doesn't notice it, the stars in our night skies are tinged with color. If you don't believe this, next time, on a good dark night, take a look at the sky. Of course, there will be stars that have no color (white stars) but shine brightly. But if you look hard enough, you will see stars that have a tinge of yellow, orange, or red. These stars are also radiators of photons. Our sun is essentially a blackbody and radiates photons as a function of its surface temperature. In fact, all stars approximate blackbody behavior; the photons that radiate from these stars do so mainly as a function of the surface temperature of the star. Stars that are really hot (> 30,000°F) appear blue, and those that are coolest (3,000°–6,000°F) appear red or orange. In-between (6,000°–17,000°F) stars are orange, yellow, or white.

Table 1.1. Temperature and the color of a star

Type	Range	Color
M	3,000°–6,000° F	Infrared/Red
G	6,000°–10,500° F	Orange/Yellow/White
F	10,500°–13,000° F	White
A	13,000°–17,500° F	White
B	17,500°–50,000° F	White/Blue
O	50,000°–100,000° F	Blue/Indigo/Violet/ Ultraviolet

White light emitted by a sun is mostly the combination of photons in the wavelength range of our visible light, from 400 nm to 700 nm. Red light emitted by a sun is the combination of photons in the wavelength above 700 nm but less than 1,000 nm. Remember, though, that other wavelengths of light are radiated by these stars but at a lower amount than the predominant kind of radiated light. Our sun radiates mostly white light (our sun is about 5,800° F) but also radiates photons with wavelengths above 700 nm and photons with wavelengths below 400 nm. About 85 percent of the photons emitted by our sun are in the visible range of 400 nm to 700 nm; the remaining 15 percent are outside of this range.

Note that one of the major colors of the spectrum is missing from the list in Table 1.1—green. Oddly enough, even though our sun looks orange in the sky and radiates photons across the visible spectrum, its peak radiation is about 480 to 500 nm. Guess what? This is actually the wavelength of green light. So why doesn't our sun look green then? In order for our sun to be a green color, the grand majority of photons would have to come from this range, but that is not the case with our sun or any other star in the universe. The photons radiating at other wavelengths wash out the green color, and we end up with a spectrum of photons that make our sun appear white. In fact, there are apparently no stars (or so very few that we have yet to see them) in the universe that can radiate green. There is just too much residual photon activity around this green peak to allow for green to leak through.

In 1953, the great science fiction writer Philip K. Dick described an encounter between humans and the native inhabitants of a planet circling Betelgeuse, a red star in the Orion constellation. This star is easily visible as the ninth brightest star in the night sky, and it radiates red light even to the naked eye. At about 600 light-years away, Betelgeuse radiates light in mostly the infrared range with a peak radiation of 750 nm. Dick's short story "Tony and the Beetles" is more about xenophobia than anything. It's a story about a human boy befriending Betelgeuse natives in the town Karnet on a planet revolving around Betelgeuse. They're facing an interstellar war between humanity and the planet's beetle-like inhabitants, the Pas-udeti. While astronomers have not identified planets currently circling Betelgeuse, there might have been, a long time ago before this star started to expand. But Dick did his homework (as he always does) about what vision would be like on a planet circling Betelgeuse. He starts the story with the following: "Reddish-yellow sunlight filtered through the thick quartz windows into the sleep-compartment." Indeed, such a planet would be inundated with near infrared and red light. Tony, the young protagonist of the story, would see things in shades of red because the majority of photons hitting his human eyes would be reflecting off of objects and would be in the 700–750 nm range. If Tony could wear goggles that filtered out light of wavelengths greater than 700 nm, he would probably see color with his human eyes, but it would be a very dimly colored world he would visualize. The amount of photons in the range of 400 to 600 nm would be very low, leaving fewer and fewer photons of the correct wavelength to hit his eyes.

But remember that Tony is a human transplanted to this planet circling Betelgeuse. What about the Pas-udeti, those beetle-like organisms that he befriends? Well, Dick doesn't say much about the biology of these organisms, other than they are beetle-like and pissed off about human domination in their own star system, so we can use a little poetic license here. Let's assume that detecting more than just visible red light was of selective significance to the Pas-udeti, say to recognize castes or even to recognize sexes. More than likely the Pas-udeti would have evolved a very different manner of sorting out wavelengths of photons hitting their visual systems and perhaps would have evolved a system of "color" vision that could distinguish between various wavelengths of red, near infrared, and infrared, some in the range of wavelengths our eyes do not detect. As we pointed out earlier, there are optical devices that allow the visualization of light in this infrared range, and indeed there are organisms on our planet that can discriminate photons therein, which we will talk about later in this book. Since we don't know much about the ancestors of the Pas-udeti, there is no way to make a statement about whether the evolutionary solution for color vision of this organism is close to perfection, but the best bet would be that it would not be. If Tony's descendants and the descendants of humans on this planet don't get decimated by the angry Pas-udeti, then these human populations will more than likely evolve with respect to color vision, and since we know something about human color vision, we could say something about continued pressure from infrared light photons on how the human color vision system might adapt to Betelgeuse's photon radiation.

Because we know the ancestral state of how humans process photon information, we can try to predict the kinds of changes that might occur as human color vision systems are exposed to mostly near infrared and infrared light by living in the Betelgeuse system. The existing color vision system of humans on Earth would be what natural selection would probably work on. More than likely, the human color vision system would evolve toward being able to interpret more information, and hence to process the near infrared and infrared photons hitting human eyes. Or perhaps humans might evolve another organ to detect infrared and near infrared wavelengths. Eyes aren't the only things that can detect electromagnetic radiation in organisms on our planet, as we will see later in this book. Either way, after a significant period, the humans there would be adapted to the > 700 nm photons and more than likely lose the capacity to process photons in the 400- to 700-nm range. If there are no or few photons of this range of wavelengths radiating to the surface of the planet, then it makes sense that it would lead to loss of function of the old system, which takes up a lot of energy and time processing photons of lower wavelength. Humans living in the Betelgeuse system would see the universe in different wavelengths than humans who remained on Earth. Transplanting such a Betelgeuse human to Earth would result in a human who could detect photons like the Predator we discussed earlier, but also with very poor vision in our visible spectrum.

Other evolutionary possibilities might exist for our Betelgeuse humans, because the actual mechanism by which color vision works in humans on Earth is based on molecular biology and biochemistry. They might even evolve to use a completely different molecular basis for color detection. How this might happen is the next part of the story of the natural history of color.

2

Color without Eyes

If we say something like "Light is perceived by a broad range of organisms," what exactly do we mean? The perception of light by us humans is one thing, but do microbes perceive light? Do plants? Do things without eyes? The key word in the sentence above is *perceive*. Perception is usually connected to things with brains and nervous systems. Some scientists have suggested that plants have nervous systems, or something like it. The journal *Plant Signaling and Behavior* was started in 2006 to focus on some of the peculiar behaviors of plants and the possibility that plants have a nervous system. Having a nervous system and having "something like a nervous system," though, are two completely different things. The establishment of a new journal focused on a specific topic usually means that a new discipline has arrived. Indeed, researchers have focused a lot of attention on the subject of plant behavior and "neurobiology," and not without controversy.

Two camps immediately formed—one convinced that plants have nervous systems, the other best characterized by the title of a news article, "Intelligent Plants or Stupid Experiments." Let the butting of heads begin, with calls for retraction of papers based on poorly conducted experiments and a grain of general ill will.

The Difference Between "Having a Nervous System" and "Having Something Like a Nervous System"

Some researchers hedge their bets and simply suggest that terms like *plant brains* or plant *nervous systems* or the products of such, like memory and learning, are simply metaphors. Others, like Monica Gagliano, a behavioral ecologist in Australia, are much more adamant. "My work is not about metaphors at all. When I talk about learning, I *mean* learning. When I talk about memory, I *mean* memory," she says. She echoes the sentiments of two of the plant neurobiology field's founders, Frantisek Baluska of the University of Bonn and Stefano Mancuso of the University of Florence, who vociferously maintain that nervous systems *literally* exist in plants. Clearly there is a problem here. But it is a problem as old as the process of comparing things in human history. The problem simply stated is when do we know two structures in different organisms are the same thing? Aristotle faced the problem in his writings and botched it as he oftentimes confused structures in one organism as existing in others. A comparative anatomist named Pierre Belon got the right answer, but for the wrong reason, in the 16th century. Belon is now famous for a single figure he drew in 1555 of a bird and a human skeleton next to each other, pointing to all of the bones in one that he thought were in the other, and he was, in general, correct. The natural philosophers who came to prominence in the 17th and 18th centuries also had the problem (and mostly botched it). Darwin (who finally got it right) faced the problem in formulating his game-changing ideas about natural selection and diversity of organisms.

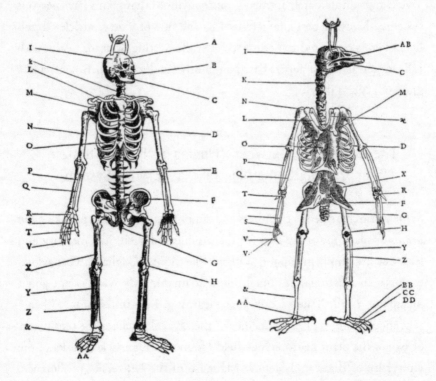

Figure 2.1. Belon's homology drawing. The twenty-seven bones of the human skeleton (left) are homologized to the bones in the bird skeleton (right). While Belon had little interest in evolution and no formal understanding of common ancesty, he was still able to homologize the structures in the figure. *Drawing in public domain.*

Darwin, by the way, also wrote about plant brains in *The Power of Movement in Plants* (1880), where he wrote about the radicle, the embryonic root in a plant, and the sensitivity of its tip to diverse kinds of stimulation. Darwin thought that perhaps the radicle "acts like the brain of one of the lower animals." Not only does the radicle act like a brain but it is also anatomically positioned at the anterior end of the plant, as brains are positioned in animals. But Darwin hedged his bets with the words "acts like"; he knew better than to make the claim of "plant something" equals "animal brain."

To humans, metaphors are incredibly important. They help us understand difficult, weird observations and expand our approaches to the discovery of novelty through comparison. Metaphors are only important, though, when they are recognized as such. We sadly doubt that the memory and learning that Gagliano claims to be so like our general understanding of learning and memory in animals, as well as her literal interpretation of plants having nervous systems, is the correct way to conduct research. Why make such a stink about metaphors? What we are really talking about here is a comparative concept called "homology," which the famous evolutionary biologist J. Maynard Smith called "a term ripe for burning." One person's bane is another person's raison d'etre, though. As Gary Nelson, a less famous but equally important evolutionary biologist, once said, "Nothing in evolution makes sense except in the light of homology." He was of course building off another famous evolutionary biologist, Theodosius Dobzhansky, who once said, "Nothing in biology makes sense except in the light of evolution." Smith rejected homology as a useful term, and Nelson lived by it. As Darwin explained, structures are homologies if they exist in organisms because of common ancestry. They are analogies if they occur as a result of convergence. It is important for our discussion of color because, when we attempt to say that color to a shrimp is the same thing as color to a chimpanzee, we are assessing the homology of the concept of color in both lineages.

Plants do indeed respond to and utilize light. Sunflowers are particularly adept at sensing light, as they follow the sun across the sky by positioning their flower head directly with the rays of the sun, a phenomenon called "heliotropism." It should be noted that only very young developing plants do this (older plants point stably eastward). The plants are using circadian rhythm mechanisms dictated by genes in their genomes. Not all heliotropic plants accomplish their sun seeking in the same way; in other words, not all mechanisms of heliotropism are homologous. Sunflowers use auxins, small molecules that are hormones coursing throughout the plant. They are important in gene regulation for the developing and growing plant. The auxin hormones are growth hormones and themselves are everything but heliotropic.

In fact, the auxins shun light by moving to areas of the plant where sunlight is shielded. The auxins stimulate growth in the regions of the plant where they end up. Since growth occurs in the shady region of the stem, the plant gets chubby in that region, and the chubbiness in the shady region bends the plant away from the chubby part of the stem—that is, away from the shady part of the stem toward the sun. Because the Earth is rotating (and the sun seemingly moving across the summer sky), the auxins scramble to get away from the sun and in so doing make other regions of the stem chubby relative to each other. Again, the chubbiness bends the plant toward the sun.

It might at first seem that the plant is behaving like an animal moving to collect as much of the sun as possible (or, as some animals do, seek shade from the sun). However, the movement of the plant in response to the sun couldn't be more different from the movement of an animal in response to the sun. Some heliotropic plants, like sunflowers, follow the sun from east to west during the day, then return their flowering heads to the east at night, as if anticipating the sun rising in the east. The simple inference is that they remember that the sun starts to rise in the east and so they try to get a head start on soaking up the rays. Memory has little to do with it, though, as the molecular workings of circadian rhythm are directing the west side of the plant at night to grow and get chubbier, with the flower head turning because the stem gets thicker. One could say that circadian rhythm is a form of memory, but memory in animals is much more complicated than any circadian "memory," replete with storage in a brain and a slew of molecular interactions much more complex than the auxins and reacting much quicker than the system that implements cell growth in sun-tracking flower heads.

One question we haven't answered with respect to our sun-loving sunflowers is, Why do they use this unique way of maximizing the amount of light they are exposed to? The answer to this question gets us closer to how organisms that don't perceive light with a brain deal with light. Plants of course use sunlight (the photons from sunlight) as food, so to speak. It is where they get their energy and where storage of molecules, like sugars, start. But before we can delve into plant detection and utilization of light, we need to dig much deeper into the history of life on our planet to understand where the plant way of life with light came from.

Life in Hell

The single most significant kind of organism that ever lived on this planet is not us. Rather, it is more than likely a group of tiny bacteria called "cyanobacteria." To see why, imagine our planet 3.5 billion years ago. Needless to say, it was a very different planet than it is now and indeed very different from even one billion years later. Prior to 2011, most scientists were convinced that the Earth's atmosphere was the product of massive volcanic activity on the forming planet's surface. Such an atmosphere would have created a highly reduced wasteland full of methane (CH_3) and other equally nasty compounds, like carbon monoxide (CO), hydrogen sulfide (HS_2), and ammonia (NH_3). This highly anoxic, or reducing, atmosphere was supposed to be a major stumbling block to life as we know it. Why? Hardly any oxygen in it at all. The party line was that a process we will shortly discuss converted all of this bad stuff into compounds that life as we know it could evolve in, and then it was off to the races. From this poisonous Garden of Eden it took a billion years to evolve into the oxygen-rich hospitable planet we now have.

American astrobiologists Dustin Trail, Nicholas Tailby, and Bruce Watson examined rocks from the Hadean eon, aptly named because, as you might guess, it was a hellish time on Earth. Their work, though, showed that the Earth was not as noxious as hell but rather contained compounds like water (H_2O), carbon dioxide (CO_2) and sulfur dioxide (SO_2). All, as the astute reader can see, contain oxygen atoms. The Hadean was not as hellish as its name implies. Researchers preferred this version of early Earth because it made explaining how the building blocks of life, like amino acids and nucleic acids, might have arisen out of compounds like methane and ammonia. Instead Watson and his colleagues argue that because the atmosphere was most likely not chockful of these life-essential building block precursors, the building blocks perhaps came from somewhere other than our planet. But we have digressed a bit, because our focus here is on how sunlight and more specifically photons are perceived by organisms on our planet. It turns out that this reimaging of the hellish atmosphere of the early Earth isn't inconsistent with how the atmosphere became as highly oxygenated as it is today. In that respect, it doesn't interfere with our story of how light and eventually colors began to be used, and in the end perceived by organisms on our planet.

What was still needed was something that could trigger the oxygenation of the atmosphere, or, as it is known, the Great Oxygenation Event. That something it turns out was and continues to be the single most influential organism on this planet, a small kind of microbe called "cyanobacteria." Most bacteria, as well as fungi and most of us eukaryotes, are what we call *heterotrophic*; we suck in or ingest compounds to make a living. Cyanobacteria, a single-celled organism considered to be a bacterium, evolved a mechanism whereby it gains its energy and makes its living through sunlight. This non-heterotrophic bacteria evolved internal membranes called "thylakoids," where the special oxygen conversion occurs. This process is called "photosynthesis." As its name implies, it takes light (*photos*), water, and carbon dioxide and converts (*synthesis*) them into nutrition and energy for organisms that use the process to make a living.

Photosynthesis in cyanobacteria is a straightforward reaction. Those readers who are not chemically inclined might need to skip over the next few hundred words, but suffice it to say, it really is pretty simple chemistry. Remember that nature is the ultimate bookkeeper when it comes to chemical reactions. Only rarely do the books get tipped, and the broad majority of the time the books are intricately balanced. We start out with six molecules of carbon dioxide (CO_2, or one carbon and two oxygens), as well as six molecules of water (good old H_2O, or two hydrogens and one oxygen). Sunlight then catalyzes the reaction of these twelve molecules into a single molecule of sugar ($C_6H_{12}O_6$ and six molecules of O_2). We won't go into why the books are kept balanced (it has everything to do with the atomic structure—specifically, the electrons of the participants in the molecules), but we will make sure that the books are balanced. The equation looks like this:

$$6CO_2 + 6H_2O \longrightarrow C_6H_{12}O_6 + 6O_2$$

The arrow indicates sunlight catalyzing the reaction. There are six carbons, eighteen oxygens, and twelve hydrogens on the left and—as the Violent Femmes song says, "Add it up"—there are six carbons, eighteen oxygens and twelve hydrogens on the right side. The process uses up noxious CO_2, produces sugar for nutrition, and gives off oxygen into the atmosphere, completing the ultimate triple play in nature. What we are after, though, is what the light does, and that is where we get back on track with our story of organismal perception of light.

Cyanobacteria live in water, so the water on the left-hand side of the equation is plentiful. They also prefer nutrient-rich water, and, you guessed it, CO_2 is in nutrient-rich water and in the atmosphere (unfortunately in higher and higher quantities as a result of human misuse of the environment), so no problem with the left-hand side of the equation. The right side of the equation is no problem either, and in fact, the raison d'etre of photosynthesis is the right-hand side. It's that arrow in the middle of the equation that we now turn to. Explaining it will tell us how cyanobacteria and other organisms that use light get this reaction to work so well.

As we saw in chapter 1, light is both a particle and a wave. The particle nature of light is important in how the arrow in the above equation works. Sunlight comes into the cyanobacteria. It can either go straight through the cell, be reflected off something in the cell, or be absorbed by the cell. It is this last function that is important for photosynthesis. When the sunlight gets absorbed, it has to do something to the cell, as the laws of physics tell us. Cyanobacteria have small molecules called "pigments," which absorb sunlight and store the energy from photons that are then used to catalyze the reaction we have been focused on. The major pigment that does this is called "chlorophyll." This category of molecules has a very important characteristic that runs part of the photosynthesis show. Its molecular structure includes what is called a "porphyrin ring." Chemical rings are pretty amazing structures; this one is relatively stable but allows electrons to move freely around it. If something interacts with the chlorophyll (like a photon), the chlorophyll molecule gets excited and starts the electrons moving around, making them available to go about their reducing ways. In essence, the chlorophyll molecules lose electrons from their porphyrin ring, making them available to other molecules nearby, and in turn available to participate in chemical reactions like the one in the equation above. Chlorophyll doesn't necessarily "absorb" or capture the photons of light.

Remember, light is particulate but not really a particle, and it is wavelike but not really a wave. The photon essentially loses its energy and its capacity to move in space, and hence appears to be absorbed or captured. A side product of this "capture" is that the wavelength of light captured is absorbed. So what happens is that all wavelengths of light that don't go through the cyanobacteria or get captured by chlorophyll get reflected off the bacteria. Cyanobacteria have evolved two major kinds of chlorophyll molecules, perhaps unimaginatively named "chlorophyll a" and "chlorophyll b." It turns out that these chlorophyll molecules are particularly good at capturing light in the blue (around 430 nm) and red (around 662 nm) wavelength ranges, reflecting, you guessed it, photons of green (around 560 nm) wavelength. Which, as you might have guessed, is why the molecules are called "chlorophylls" (*chloro* meaning "green" plus *phyl* meaning "leaf") and why cyanobacteria are called such (*cyano* meaning "blue green"). They are also known as blue-green algae, but this is a misnomer not because they aren't blue green but because they aren't technically algae.

But we are getting a little ahead of ourselves. When cyanobacteria evolved photosynthesis, there were no animal eyes around to process the reflected green, let alone the reflected green-blue wavelength photons. Why do chlorophylls capture red and blue wavelength photons? One reason might be the "more the merrier" rule and the idea that these two kinds of chlorophylls maximize the energy captured from sunlight. Cyanobacteria have other pigments that focus on capturing photons of wavelengths chlorophylls avoid. Carotenoids are a category of pigments that capture green-blue wavelength photons between 450 nm and 475 nm and most efficiently reflect light in the red-orange range. These pigments cannot directly transfer electrons, though, so they are accessory pigments. A third category of pigments in cyanobacteria is the phycobillins, and these most efficiently absorb light at wavelengths at about 570 nm (or light of green wavelengths). Phycobillins usually reflect reddish-orange or brown light.

Sounds a little like cyanobacteria are sensing color, doesn't it? If they have molecules that are specific for these wavelengths, then they are in a way perceiving light of different wavelengths, right? Wrong! Cyanobacteria are not processing the photons that hit their pigment molecules in the context of what the light means, nor do they care what it looks like. They are using the light of different wavelengths as efficiently as they can for energy to drive critical reactions in their cells. Cyanobacteria are terribly pragmatic when it comes to how they use light of different wavelengths. It is this basic cellular pragmatism that sets how this bacterium uses light apart from other organisms.

Blind microbes?

So far, we have been discussing cyanobacteria and photosynthesis. But light is used by other single-celled organisms for similar essential, life-giving functions. Depending on its overall lifestyle, a single-celled organism sometimes needs to pump stuff into and out of its interior. Usually in doing so, energy is used and some chemical reaction that is essential to the life of the cell is accomplished. The energy to do this again comes from sunlight. One of the more common mechanisms some single-celled organisms need to perform (and indeed cells of multicelled organisms, as we will soon see) is to pump protons (positively charged particles in the nucleus of the atom) into and out of the cell to balance the charge or to create charge for the cell.

As with chlorophyll and cyanobacteria, a molecule does the job. But this time the molecule is actually a protein, and instead of being glued to small membranes sitting inside the cell, it does its job embedded in the outer membrane of the cell. "Proton pumps," as they are called, are so important that nature has figured out how to make them independently many times. What this means is that not all protein pumps are related to each other through a single common ancestor. Yep, we are back to homology as a guiding principle for how we view innovation in nature. While the origins of the proton pump systems are independent of each other, they operate on the same principle, which is to use some source of energy to implement the movement of protons across an otherwise non-impenetrable membrane.

Bacterial rhodopsins, also known as "retinal proteins," are important proteins that weave their way from the outside of the cell through its membrane seven times, so that the other end of the protein ends up on the inside of the cell. They exist in what we have so far been calling "bacteria," which require an intricate balance of charge on the inside and outside. Until about the 1970s, people regarded all life on the planet as falling into either of two large impressive domains—Prokaryota and Eukarya. This meant that there were two basic kinds of cells on Earth. The eukaryotes, both single-celled and multicellular, had cell membranes and a nucleus surrounded by a membrane, where the genetic material resided. The prokaryotes, universally single celled, were organisms with a cellular membrane but no nuclear membrane, so that their genetic material swam around in the cell, unprotected.

Enter Carl Woese, a microbiologist from Illinois. In the 1970s, he used the intuition of two giants of science, Emile Zuckerkandl and Linus Pauling, to look closer at this major division of life. In a series of wonderful papers written in the early 1960s, Zuckerkandl and Pauling outlined something most biologists today take for granted. In their 1965 paper entitled "Molecules as Documents of Evolutionary History" (actually first written in 1963 and published in English two years later), they suggested that a molecule, like DNA, since it is inherited from parent to offspring, carries information about the relatedness of organisms. As mutations occur in the DNA of a parent organism, it is passed on to the offspring organism, and so and so on. Since proteins are coded for by DNA, any changes in the genes of an organism will also show up in the proteins of organisms. These changes could indicate two things. First, the more shared by organisms, the more closely related they will be. And second, more changes between organisms might also indicate distant relatedness.

Woese's innovation was to use this idea to look at differences between the single-celled organisms that were so important to him. He and the scientific community were shocked by the results. Instead of only two domains of life, there are actually three. Discovering a new species, or even a new genera or family, is a feat of scientific accomplishment, but a whole domain—Oh my! Woese split the prokaryotes into bacteria and archaea and left the eukarya alone. More interestingly (what could be more interesting than discovering a new domain of life?), he and his colleagues eventually showed that archaea are more closely related to eukarya than to bacteria. Let's be clear here though; before Woese, researchers knew that there were lots of bizarre kinds of single-celled organisms they called "prokaryotes." Bizarre because they liked to live in extremely inhospitable environments like hot springs, hypersaline environments like the Great Salt Lake, and deep in the ocean. Hence, they were called "extremophiles" (lovers of extremes). They were also sometimes called "archaea" because researchers thought they were the archaic or primitive forms of microbes that lived on the planet 2.5 billion years ago. Woese realigned this thinking completely.

Halophilic archaea have a special need. Because they live in hypersaline environments, their ionic milieu is somewhat complicated. In essence, to fill that hypersaline niche they needed to figure out a way to regulate ionic conditions and utilize this scarce resource. Proton pumps were the best way to do this, and their rhodopsins, more appropriately called "bacteriorhodopsins," were the best way to make the pump. How they work probably won't surprise you. As they sit in the membrane with one end dangling outside the cell, laced through the membrane seven times, and coming out on the inside, they wait for light to hit them. When light hits the rhodopsins, they change shape. This shape shifting triggers a series of reactions inside the cell that make an important molecule called "ATP," which is used for energy. Rhodopsin gets most excited by light in the green range, with its sweet spot being 568 nm wavelength light. So, it absorbs green light and reflects red light most efficiently.

But the reflected red light looks a little red-purplish, and halophilic archaea are often said to have purple membranes. Some of the early experiments with halophilic archaea that attempted to decipher the role of these important membrane proteins took strains of a species called *Halobacterium salinarium* that had mutated and lost all of its rhodopsins. Sergei I. Bibikov and colleagues call it a "blind" halobacterium. Granted, it could not utilize light, and to Bibikov's credit he used quote marks around the word *blind*, but here again is metaphor horning in on the show. This case is pretty clear cut. The halophiles are not blind in any sense of the word. They simply lack proteins that some close relatives have, and instead of lacking sight, they lack the capacity to absorb green light and pump protons with rhodopsins. It's as simple as that.

It should be pretty evident that light is essential to organisms and life on this planet, and we haven't even made it to the so-called more complex eukaryotes. When we do consider eukaryotes, we see some of the same general themes as to how these organisms deal with light. In fact, it shouldn't be too surprising that eukaryotes have some of the same mechanisms and even molecules that single-celled bacteria and archaea have. After all, these organisms without nuclei are related by common ancestry to organisms with nuclei. But before we can talk about these major steps in eukaryotic cell evolution, we have to talk a little about eukaryotes themselves and how diverse they are.

Games of Thrones

The late Lynn Margulis, who was described during her amazingly productive career in biology as "science's unruly earth mother," divided the world into five kingdoms—Monera (basically prokaryotes, archaea, and bacteria); Protists (single-celled eukaryotes); Plants; Animals; and Fungi. Taking the drastic step of defining five kingdoms was very unruly indeed, but it wasn't even her most disruptive idea as a biologist, as we will soon see when we talk about chloroplasts and mitochondria. This five-kingdom system was a great heuristic way to organize life for many reasons. It was simple and at the same time elegantly based on some of the known characteristics of organisms. But it failed in a couple of ways that are near and dear to the scientists who practice taxonomy and systematics. The utility of these two disciplines of biology is that scientists systematize and name groups of organisms that they can recognize in nature. A common guiding principle of these two disciplines is a concept called "monophyly." This strange little word has as much baggage as the term "homology" that we explored earlier in this chapter. But it is an important concept in organizing things in nature to make it easier for the human mind to process what the diversity around us means. A group of taxa are monophyletic when they all have a common ancestor to the exclusion of other taxa. A simple example would be the statement, "Eukarya is monophyletic relative to Archaea and Bacteria." This is a true statement because all eukaryotes have a single common ancestor that excludes archaea and bacteria. Monophyly is a great principle because it boils relationships of organisms down to a single objective criterion—inclusion in a group to the exclusion of other things. Monophyly operates on all kinds of levels, and this is where the term gets a little confusing. Now let's look at Margulis's nomenclatural mess-up.

As we saw above, eukaryotes are a good monophyletic group, as are bacteria by themselves and archaea by themselves. But prokaryotes are not. Why? Because eukaryotes and archaea appear to have a more recent common ancestor than bacteria and archaea. Prokaryotes imply that archaea and bacteria have a more recent common ancestor and exclude eukaryotes, which simply isn't true. So there goes Monera and prokaryotes with it. Fungi, Animals, and Plants have no problems with monophyly, as they are all pretty good monophyletic groups. But it's those pesky protists that show Margulis's second nomenclatural mess-up. Protists are amazingly diverse kinds of organisms. Notice we didn't say "group" of organisms, as they aren't really a group in the monophyletic sense. This non-groupness of protists is caused by the fact that many of these single-celled eukaryotes have affinities for plants *and* animals. So, for instance, the protist algae have greater affinity for plants than they do for animals, and the amoebal protists have stronger affinity for fungi and animals than for plants, and so on and so on for other protists. So, protists aren't a monophyletic group at all, messing things up royally in the nomenclatural world of Margulis's five kingdoms. What to do then? Well, the most precise thing to do is to call the protists their most inclusive and simplest-to-understand name, like algae or amoeba, and only use protist when precision is not required. This is probably a very dissatisfying solution for many readers (as it is for us too), but it is the best we can do without naming each major protist lineage a kingdom, raising up tens of new kingdoms, creating a massive game of thrones with far more than seven kingdoms (or six if you count out the North). Why is this all so important for understanding color? Let's go back to our goal of understanding how light and eventually color is perceived. If we muddle taxa and make poor sense of their evolutionary history, then we will have difficulty making sense of the diversity of mechanisms that have evolved for light perception.

Dog Eat Dog

The majority of eukaryotic cells have relatively small organelles called "mitochondria." Some highly specialized eukaryotic cells like our very own red blood cells have managed to rid themselves of mitochondria during part of their developmental pathway. However, suffice it to say that mitochondria are essential for at least part of these cells' lifestyle and hence essential to energy production for eukaryotic cells. These small, self-contained entities reside in the cytoplasm of the cell and perform the essential job of providing energy for the cell to live. Mitochondria have membranes and their own small, circular piece of DNA with thirteen protein-coding genes, two genes that make relatively long RNA molecules that fit into ribosomes, and a complement of about twenty smaller RNAs called "t-RNAs." The thirteen protein-coding genes code for enzymes that participate in the reduction of molecules from the food the organism eats. The enzymes reduce oxygen within the mitochondrion to produce energy for the cell. This small number of genes in the mitochondrial genome means that mitochondria borrow genes and proteins from the eukaryotic nuclear genome to carry on their functions; in this sense mitochondria could not function outside the eukaryotic cell. While they have membranes that encircle them, they are by no means capable of independent life.

But mitochondria don't use light to produce energy, like cyanobacteria. They use other sources to generate the essential needed energy—oxygen and the chemicals from the food the organism ingests—that the cell needs to carry on its lifestyle. Another cellular organelle called a "chloroplast" is found exclusively in photosynthesizing eukaryotes. It also has an outer membrane and a genome like mitochondria but is much larger with over 100 genes, many of them involved in processing light into energy. The chloroplast is so named because it emanates a greenish color as a result of being packed with, you guessed it, chlorophyll. The chlorophyll clings to small membranes internal to the outer chloroplast membrane, called "thy-lakoiods." Wait, sound familiar?

The question here is, Did eukaryotic cells evolve mitochondria from scratch, and did photosynthesizing eukaryotes do the same for chloroplasts? The answer is No, but rather that a shortcut was taken where the eukaryotic cells kidnapped other already existing cells they were in contact with. These kidnapped cells then accomplished the job of energy production and light processing. Two of the most important events in the evolution of eukaryotic life on this planet were microbial kidnaps, or, more accurately, microbial culinary events, rather than the result of good old, missionary style common ancestry. Our cells, like the cells of microbes, need to manage ions and energy and all kinds of other important life-sustaining processes. They could have evolved these mechanisms via good old mutation, drift, and natural selection over eons of time, but eukaryotes took two serious shortcuts, one that gave us cells that can generate their own energy, which all eukaryotes have, and another that gave us eukaryotes that can photosynthesize.

The term "dog eat dog" doesn't have a thing on "microbe eat microbe." Eating is not really the most accurate way to describe what happened; perhaps a better description is a mutually beneficial love story between two kinds of cells that ended in everlasting bliss. However, preferring a good zombie story like "The Walking Dead" to a love story like *The Notebook* precludes us from metaphorizing things with the love story. Instead, let's put it the way Lynn Margulis did when she first popularized the idea of engulfment of microbial cells by the ancestral eukaryotic cells. In the 1880s, Andreas Schimper first came up with the idea of plastids being endosymbiotic. He was followed in this idea by a Russian botanist named Constantin Mereschkowsky; both seem to have gotten lost on the trash heap of history, although their names are mentioned occasionally (Margulis, to her credit, cites Mereschkowsky in her 1967 paper). Margulis explains the idea this way: "The mitochondria, the basal bodies of the flagella, and the photosynthetic plastids can all be considered to have derived from free-living cells, and the eukaryotic cell is the result of the evolution of ancient symbioses." This paper, by the way, was apparently rejected from fifteen journals, and her work on this subject was considered pretty "crappy" by most biologists. She even admitted she didn't know much about the organisms but that the obviousness of the idea was clear to her. (At one point in this remarkable paper she mentions that "it is likely that the classifications presented in the phylogenetic tree err in that the author [herself] lacks first-hand knowledge of most of the organisms.") While it is incredibly difficult to wade through, the paper has now been cited over twelve hundred times and is considered a classic in the scientific literature. Things just looked right to the unruly earth mother, and, being the contrarian she was, she gladly took on the scientific community and defended the idea that mitochondria come from a bacterial precursor and chloroplasts come from a cyanobacterial precursor cell. While Margulis supported her ideas based on biological inference, many researchers poo-pooed her and ignored her ideas for some time. Two things shoring up the endosymbiosis idea have happened since her publishing the 1967 paper; one a little old and one a bit newer.

The chloroplast and mitochondria look like, and have qualities like, bacteria. It wasn't until researchers could sequence the small genomes inside these organelles and compare them to existing bacterial sequences, like Carl Woese did in the 1980s, that the mystery was solved. The approach Woese used is a little like fishing. You throw the organismal sequence data out as bait into a pool of chloroplast or mitochondrial sequence data, and where the chloroplast and mitochondrial sequence data stick is where they have common ancestry. If chloroplasts have independent origins and are not the remnants of an endosymbiotic relationship, they shouldn't stick to any one microbial taxon. Likewise, if mitochondria have independent origins and aren't the result of an endosymbiosis, then they should also not stick to any particular microbial taxon as a result of common ancestry. When this fishing expedition is conducted, it is clear that chloroplasts take up the cyanobacterial bait and mitochondria take up the bait from a group of bacteria called "alpha proteobacteria." Not many of the members of this group should be too familiar to readers, as they are a strange group of bacteria, but one member, *Rickettsia*, causes several sicknesses in humans, like Q fever and scrub typhus. Using this same approach, researchers have more recently tried to pin down a more precise timing of the chloroplast endosymbiosis event by offering more cyanobacteria bait to the experiment. Cyanobacteria are a pretty diverse group (current estimates are from three thousand to six thousand, but it is certainly much larger), and saying that cyanobacteria gave rise to chloroplasts is like saying that dinosaurs gave rise to birds. The idea is that one particular lineage of cyanobacteria (like one particular lineage of dinosaurs) might have given rise to chloroplasts. While this line of investigation is preliminary, the early returns indicate that two groups of cyanobacteria—Nostocales or Chroococcales—could be the potential "mother of all chloroplasts."

The second important experiment in this saga comes from our American Museum of Natural History colleague Eunsoo Kim. She and Shinichiro Maruyama were able to take pictures of eukaryotic cells actually eating cyanobacteria, simulating the potential for the ancient event. Kim and Maruyama show clearly that the cyanobacteria are pulled into a mouthlike opening (metaphor avoided) and then swooshed down a gullet (apologies for the obvious metaphor) into vacuoles, where the eukaryotic cell, an alga, stores the cyanobacteria for food. Unfortunately for these cyanobacteria, they suffer a worse fate than that ancestral one that figured out how to coexist with that ancient eukaryotic cell. Pretty good proof that the "ingestion" of cyanobacteria by eukaryotic cells, especially algae, is not only possible but very probable. Now that we have made the story rather simple, we have to warn you that in actuality, as our colleague Eunsoo Kim has warned us, it is complicated. There have been multiple engulfment events in the history of algae alone, and even some cases of engulfment followed by something engulfing the first diner. The algal anthem seems to have been "Eat AND be Eaten." Even with all of this eating going on, algae are still green because of chloroplasts; at least some of them. There are also red and brown algae, but these actually have chloroplasts too and use chlorophyll and absorb blue and red light. They are also chockful of pigment proteins that obscure the green and reflect red wavelength light, making them both brown and red. It is rather interesting to learn that the plants on this planet that give us so much joy because of their greenness are so because of a rather cannibalistic origin.

Plants have taken the use of sunlight to bigger and better levels than bacteria and algae have. Plants have evolved three rather different ways of photosynthesizing and fixing carbon—detecting and utilizing sunlight to concentrate, or "fix," carbon in their cells for use as nutrition. Most plants (90 percent) are what are called "C3" plants. C3 simply refers to the fact that the product of this kind of photosynthesis is a molecule with three carbon atoms (3-phosphoglycerate: $C3H7O7P$). Researchers suggest that

C3 plants evolved first, with the other two kinds of carbon fixation being derived from the C3 mode. The second kind of photosynthesis is C4 and happens in about 3 percent of all plants. All are flowering plants. As the name implies, it results in molecules with four fixed carbon atoms (oxaloacetic acid: $HO_2CC(O)CH_2CO_2H$). The third kind of carbon fixing photosynthesizing system is Crassulacean Acid Metabolism (CAM) and evolved in a wide range of plants in different groups, including ferns, gymnosperms, and mosses. It is a photosynthetic process that plants living in dry areas have adapted.

These three mechanisms make little phylogenetic sense as they pop up all over the plant tree of life, indicating that once you have basic photosynthesis, nature can tweak it, but not a lot. If nature could really play blind watchmaker without limits, then there would certainly be more than three kinds of carbon fixation as a result of photosynthesis. It appears that these three mechanisms—C3, C4, and CAM—are all that are needed to give plants the flexibility to utilize sunlight for food and energy across a wide range of environmental conditions. Plants have also evolved a bewildering array of pigments to maximize the "capture" of sunlight across as many wavelengths as possible. Remember that chlorophyll is most efficient at red and blue light wavelengths; in order to get the stuff in between, other pigments are used by plants. These pigments give plants a wonderful rainbow of colors, which we will discuss further in the next chapter. What plants are doing is attempting to capture as much light at a specific wavelength as possible, whether it be C3, C4, or CAM, or through a wide array of pigments. They aren't remembering light or learning from it but rather using it in a very basic way. This use of light results in some interesting things that plants, algae, bacteria, and archaea can do, but in no way are these organisms perceiving light and colors the same way we do. They are much more pragmatic and mechanical about sunlight than animals are. Why say such a harsh, cold thing about our green friends? Because it helps us understand the great divide between how these organisms utilize sunlight, hence light of different wavelengths, and how animals do.

With respect to sunlight, animals—like plants, algae, and bacteria—solve the problem of light detection and collection in many different ways. Plants have their C3, C4, and CAM approaches to sunlight, as well as their many pigments. Animals have evolved tens of ways of detecting and processing light that hits them. When it comes to us humans, we are of course talking about our eyes. And we could say that animals have evolved eyes to do this important light detection function, but we would fall directly down the metaphor rabbit hole, and a pretty deep one. Instead we probably should say that animals have light-detecting organs, a different but less deep rabbit hole than using the word *eyes*. With that metaphorical rabbit hole avoided, let's take a look at these light-detecting organs in animals, keeping in mind that plants, algae, archaea, and bacteria also have light-detecting mechanisms, and that not all such mechanisms are the same. Here we can also make an interesting distinction between *mechanism* and *organ*. Animals truly have organs (collections of tissue) that detect light. Cyanobacteria, algae, and plants have an organelle, the chloroplast, that detects light and pathways that process light. We suppose that one could say that a plant's leaf is a light-collecting organ (most photosynthesis is conducted in plant leaves), but this makes my point about eyes. Leaves are not eyes by any stretch of the imagination.

Fungal Tinker Toys

Fungi, despite being considered fairly simple, are anything but. There are two major groups of fungi, the ascomycetes and the basidiomycetes, in addition to a few smaller groups. The ascomycetes include things like yeasts. Their genomes, like the genomes of the basidiomycetes, are relatively small. The basidiomycetes are organisms like mushrooms. Fungi occupy as wide an array of ecological habitats as we humans do, so their genomes are chockful of interesting genes that have helped this interesting group of organisms cope with a wide range of environments. As we have seen with plants and algae, one environmental factor that is ubiquitous on the surface of this planet is light, so fungi have evolved many ways to cope with and co-opt light. Light has a role in both the physiological and morphological traits of fungi, and they can sense light in many ranges near ultraviolet, blue, green, red, and far red. They do this the same way that plants exploit green wavelength light—with molecules called "light receptor molecules." Plants, algae, and cyanobacteria have co-opted light into their photosynthesis systems and other energy-producing pathways, whereas fungi have pirated light away to regulate their development (a little similar to the light regulation of growth by auxins in plants) and to assist in sending signals across cell membranes critical to the fungi's growth, development, and physiology. The story of light use in fungi is more a molecular story than anything else. It is instructive because similar molecular interactions are at the heart of our color vision.

Fungal light detection and response have been the subject of scientific exploration for over a century. Daniel Trembly MacDougal summarized the work done on light and fungi prior to 1900. A fungus called *Pilobolus* seems to have been a favorite of lab scientists looking for environmental effects from chemicals and light. Experiments done on this lowly little organism indicated an intricate involvement of light in fungal growth. With the advent of easy and cheap whole genome sequencing, which has also augmented many areas of light and color vision study, as we will see, thousands of fungal species have had their whole genomes decoded. In addition, the capacity to survey how a cell expresses its genes using inexpensive techniques has also led to a more accurate view of how genes involved in light detection and co-option work in cells. It turns out that a very large proportion of the fungal genome is actually regulated by light; hence several incredibly important regulatory and biochemical pathways in fungi are under tight control of light.

When light of a specific wavelength hits a molecule like chlorophyll, or a protein that reacts to light like rhodopsin, it forces a change in the conformation of the molecule it hits. The thing that light hits can be rather small, like chlorophyll, or larger, like the rhodopsin protein, but in both a small part of their structure is the center of attention for light interaction. It usually holds an electrical charge that can be ejected upon interaction with a photon of the right wavelength. It's called the "chromophore," and it is effectively responsible for the function of the light reaction and for the color that is emitted by the tissue holding it. The rest of the protein can have other kinds of protein chunks called "domains," and these are what dictate the function of the overall entity. There are eleven categories of photoreceptors across all fungi. This doesn't mean that every fungal species has eleven, but just that as a group, eleven categories have arisen. Over the two billion years of evolution of fungi on the planet, they have evolved many different ways to detect light, and at different wavelengths. If fungi can do something like that, then other organisms can too.

Parenthetically Speaking

Eyes are arguably an invention of animals. But there are many different kinds of animals on our planet. We humans are a particular kind of animal called Bilateria, because our bodies are symmetrical on the right and left sides (mostly) when bisected down the middle. Since we have already claimed that plants don't have nervous systems (or at least ones that are evolutionarily the same as ours), we guess it isn't too much of a leap to tell you that only bilaterian animals can have nervous systems and brains like ours. If an organism doesn't have a brain to interpret signals coming to it, light perception, the way we do it, only exists in bilaterian animals. Most animals that are familiar to us are bilateral, such as fish, insects, mammals, and the like. But there are also a large number of animals lacking bilateral symmetry that we need to consider before moving on to our eyes. Many of these organisms that originated prior to the common ancestor of bilateral animals also have light reception. Which brings us to a very sticky story about the early evolution of animals.

There are five major groups of animals if one considers Bilateria as a single kind of animal. These are Porifera (sponges); Cnidaria (represented by jellyfish, hydroids, cube jellies, and corals); Bilateria (we are bilateral, so we are part of this major group); and two rather bizarre groups, Ctenophora (comb jellies) and Placozoa (pancake animals). On a very cursory examination, the Ctenophora look a little like jellyfish, so they were commonly thought to be part of the cnidarian lineage. The Placozoa are extremely simple, with only six cell types (Porifera, Cnidaria, and Ctenophora all have over ten, while Bilateria have tens of cell types), with no nervous system like the Porifera. All other animals have nervous systems of some sort. There are exactly 105 ways to arrange these five groups in a taxonomic scheme that involves discrete divergence. Along with not having a nervous system, the simplest organization might lead one to declare Placozoa as the most primitive and say that it has the ground plan that all other animals follow. But this would be getting ahead of ourselves, because the organization of divergence in the early evolution of animals has, as we said, many ways it could have proceeded. For instance, one way is to place the Porifera first, with Placozoa next, with Cnidaria and Ctenophora next with Bilateria. The esoteric method for representing this arrangement is called a "Newick tree," which isn't really a tree but a series of names with parentheses to demarcate what goes with what. So this arrangement we just saw would be (Porifera (Placozoa(((Cnidaria,Ctenophora) Bilateria)))). The number of opening parentheses needs to equal the number of closing parentheses. There are three major hypotheses as to how the animals have evolved and diverged.

(Porifera(Placozoa(((Cnidaria,Ctenophora)Bilateria))))
(Placozoa(Porifera(((Cnidaria,Ctenophora)Bilateria))))
(Ctenophora(Porifera(Placozoa((Cnidaria,Bilateria)))))

With these different hypotheses of relationships, the parentheses might start to hurt your eyes, so lets simply look at the branching patterns using what are called "phylogenetic" trees.

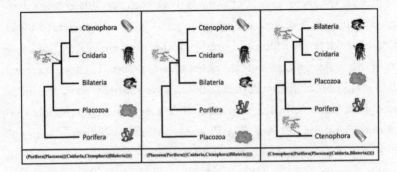

Figure 2.1. Visualizing parenthetical trees. The three trees we discuss in the text are shown here with their parenthetical expressions at the bottom of each. The neural network drawing indicates where the arisal of nervous systems needs to coccur for each hypothesis. The hypothesis on the right requires a nervous system to arise independently twice (or arise once in the common ancestor of all animals and then lost in Placozoa and Porifera). *Drawing by Rob DeSalle.*

A search of the literature reveals that of the 105 possible ways to arrange these five species, about thirty have been proposed in the literature by zoologists and molecular biologists. In other words, this is a contentious problem and one that a lot of zoologists are incredibly interested in. Why would we bring this problem up in the context of the evolution of color detection? It turns out that the very same thing that mesmerizes zoologists about this problem is relevant to our understanding of color detection. Here are three of the more popular Newick trees for this problem:

Note that each tree has a different organism that "comes" first. Each of the three trees has a different lineage coming from a common ancestor with all other organisms that follow. Each of these common ancestors indicated usually has a major anatomical innovation associated with it. The nervous system is a big question in animal evolution—after all, without a nervous system, we don't have cerebral eyes and hence cerebral color vision. The figure shows where a nervous system would need to evolve for each of the three hypotheses. The problem relevant to color detection, then, is where the brain evolves. In the left two trees, a nervous system evolves once in the common ancestor of Cnidaria, Ctenophora, and Bilateria.

So if either Placozoa or Porifera come first, there is no problem with evolving a nervous system only once. Some readers might be asking, Why should a nervous system evolve only once? This would be an excellent question, and at the heart of our homology discussions earlier. However, a good starting point for many evolutionary problems is the simplest explanation. Allowing the nervous system to evolve twice would not make William of Occam (of the famed Occam's razor) very happy. We will start with this assumption with respect to the nervous system and let other information change our minds.

The third tree requires that a nervous system evolve twice in the divergence of animals. Again, we show where in this third tree a nervous system needs to evolve. This could happen two ways. The common ancestor of all animals had a nervous system, and it was lost in both Porifera and Placozoa but retained in the common ancestor of Cnidaria and Bilateria. Or the nervous system could have evolved independently in Ctenophora and the Cnidaria-Bilateria common ancestor. Either way, with this third tree more weird things happened than in the first two. But not so weird when we look at how these animals detect light, which makes Occam's razor rather dull and allows for a lot of different ways to think about animal evolution. Still, it all means that when thinking about organisms with and without eyes, we have some animals, like Placozoa and Porifera, that categorically lack eyes and some animals, like Ctenophora and Cnidaria, that have neural tissue but arguably no eyes or at least eyes that we are familiar with.

If a sponge can sneeze, does it have a nose?

Some sponges can sneeze, albeit very slowly. This contraction and expulsion action in sponges occurs even in the absence of a nervous system and muscles. It's even more surprising that sponges also have what are called "pigment ring eyes" and behavior that responds to light, called "phototaxis."

Their eyes aren't cerebral eyes, as sponges have no brain (no nervous system, at that), but they have a molecule called "cryptochrome," which interacts with pigments. American biologist Todd Oakley and his colleagues have characterized two of these kinds of these cryptochrome proteins in a common sponge known as a "desmosponge." (There are three major kinds of sponges: desmosponges; glass sponges, also known as "hexactinellids"; and sponges with calcium carbonate "skeletons," called "calcareous sponges").

Oakley and his colleagues found that these two cryptochromes are expressed in the sponge ring eye, and that they have an uncommon co-factor molecule, which implements the cryptochrome activity and is secondarily involved in phototaxic behavior of sponge larvae. Other research has demonstrated that more than these two proteins are involved, and that these proteins are also induced to activity by light, indicating that light reception in sponges is a somewhat complex integrated system. Most of the interactions of molecules, like cryptochromes, are with opsin proteins, which we will have lots to say about in this chapter and the next. What happens near the sponge eye ring is a pretty bizarre set of protein interactions. The result is even more bizarre; Oakley and his colleagues point out they create an "aneural, opsin-less phototaxic behavior of sponges," or behavior without a nervous system but with something like a nervous system. If we go back to plants, we also have "aneural, opsin-less phototaxic behavior" too, but the molecules used have little if any resemblance to the sponge behavior.

Oddly enough, some animal groups without eyes or even light receptors can perceive light, even the incredibly simple Placozoa. This tiny, flat animal has six defined cell types (some researchers think maybe a few more) and is organized into three cell layers—to be as simple as possible, a top layer, a middle layer, and a bottom layer. It knows its top from its bottom and can also swim toward light, yet it has no nervous system or muscles. It has cilia lining its body that facilitate its movement. The genomes of placozoans contain neuropeptide precursor genes that are involved in neurosecretion in animals with full-fledged nervous systems, and these neuropeptides are distributed throughout the animal in all three cell layers. Swiss neurobiologist

Frederique Varoqueaux and colleagues examined the distribution of these neuropeptides and discovered that the signaling that these proteins do is of great importance to how its swimming behavior is mediated. They suggest that this form of signaling by these neuropeptide-like proteins in Placozoa might be the way that the ancestor of all animals implemented a primitive kind of neural communication among its tissues. This idea assumes that the mother of all animals (not the mother of all Bilateria, that would be the mother of all animals' great-great-great granddaughter) didn't have a nervous system and was like placozoans and sponges without a nervous system.

And here is where we return to opsins. Placozoans do have opsin genes, and these look a lot like another kind of receptor that animals have, called "melatonin receptors." This is another kind of protein that interacts with the hormone melatonin to regulate circadian rhythm in a lot of animals. (Many people take melatonin to set their biological clocks so they can sleep regularly.) When this placozoan opsin (called a "placopsin") is analyzed in the context of all other animal opsins, it appears to be uniquely related to all other animal opsins through the common ancestor of all other opsins. Not *the* common ancestor, but rather the first offshoot of all animal opsins. The only strange thing with placopsins is that they lack a specific element in the protein that prohibits chromophores from latching onto the molecule. The chromophore is all important to light detection, and with no chromophore, the receptor will not react to light. So, placopsins are presumed to be nonfunctional light detectors, albeit important ones. Why? Because they can act as a frame of reference for how animal opsins might have arisen.

Comb jellies, or ctenophores, have opsins and nervous tissue to boot. Their nervous tissue is arranged in what zoologists like to call a neural net without a central processing center like a brain. These remarkable animals have relatively unremarkable opsins, as the ones in comb jellies are also found in jellies, corals, hydroids, and box jellies, as well as in bilaterian animals. What makes the existence of these opsins intriguing in comb jellies and cnidarians is that both of these groups are iridescent and bioluminescent, meaning that they can produce color-enhanced effects on and in their bodies. While they don't have brains to process possible color effects, it is possible that their opsins are used to detect light produced by other individuals in their species in some way. It turns out that the opsins in cnidarians and ctenophores do a lost-and-found evolutionary dance, meaning certain opsins appear to arise and disappear in the many lineages of these animals. It is conceptually easy to lose a gene like an opsin gene, either from disuse (natural selection will allow for it to be lost in the genomes of organisms that don't use it) or through chance alone. But is it equally easy to gain an opsin gene? It just might be if the patterns of gains and losses (losts and founds) of opsins has anything to tell us. All an organism needs is a steady supply of opsin genes in its genome and every once in a while a gene will duplicate, making two copies of the gene, one of which is free to evolve away from its twin's function: voila, a new opsin gene arises.

Light-detecting organs are scattered al over the bodies of some cnidarians and ctenophores. There is a kind of jellyfish (*Tripedalia cystophora*) that actually has fairly complex light-detecting organs that very much resemble an eye. Because it has opsins too, researchers have determined that it can actually visualize shapes and shades. It actually has twenty-four light-detecting organs on its body, some of which are very simple. But there are four of these more complex light-detecting organs that actually have lenses, which always point upward toward the surface (even when the animal is swimming upside down), as Anders Garm and colleagues from the University of Copenhagen have shown. Apparently, the box jelly uses these light detectors to detect changes in light indicative of shade, and then navigates toward shady areas of the mangroves, where it lives. It turns out that many of the genes used to make these eyes are also involved in controlling eye development in higher animals. But since the box jelly doesn't have a brain, and it appears to be an isolated appearance of a light-detecting organ in this lineage, the light detectors that the box jelly has evolved are more than likely not the same thing as our eyes, or even an octopus eye or a frutifly eye. This pattern of gaining and losing light-detecting organs in animals is much like the lost-and-found nature of opsin evolution, with a twist. Instead of duplicating a gene to make a new copy for evolution to tinker with, animal genomes carry what Sean Carroll of the Howard Hughes Institute calls an evolutionary toolbox, or a set of important developmental genes that remain a steady part of animal genomes, which can be called upon to produce structures in organisms. Because the combinations of the genes in this toolbox are limited and because development of tissues proceeds in the way it does, there are a lot of structures that evolve that are repeats of previous evolutionary "experiments." We will have much more to say about this in the next chapter while discussing the evolution of bilaterian light- and color-detecting organs that have evolved over twenty-five independent times.

The evolution of light detection and the structures that do the detecting is fascinating and, in some ways, convoluted. While the genetic and molecular mechanisms that led to light detection across the animal tree of life are diverse, a general theme developed in the early evolution of animal life on our planet with respect to light detection in Bilateria. Organisms with the capacity to detect light evolved systems with a light-sensitive molecule, like a chromophore, that can bind to a protein (usually a pigment like opsin) to form a light-sensitive entity that then reacts to light, causing some other response in the animal's cells. We have already examined many of these proteins across the tree of life in organisms without nervous systems, organisms with nervous systems but very unlike ours, and organisms with nervous systems that are a lot like the precursor to our nervous system. The most prominent of these proteins that bind chromophores in organisms with eyes connected to a brain (or cerebral eyes) are the opsins. We now delve into how opsins led to and rule color vision in higher animals.

3

Color with Eyes

The famous biologist Ernst Mayr once argued that animals evolved eyes at least twenty-five to forty independent times on this planet. With our Metaphor Warning sign out, we should recognize the possibility that none of these independently evolved eyes are the same thing. They can't be if we apply our principle of homology to understanding these structures. Instead, each time a novel or independent iteration of an "eye" evolves, it is actually something else. Even though it might look like an eye, function like an eye, and use many of the same genes that most eyes do for its function, they are not all technically eyes. This is an evolutionary argument and not one based on development or genetics. There is only one kind of light-sensing organ that we could call an eye, and that is an arbitrary choice we can make. We can call the first light-sensing organ in animals "the eye," and all other eyelike things that evolved independently have to be called something else. Or we can call our human light-sensing organ "the eye," and again all other nonhomologous light-sensing organs something else. One can use the word *eye* with a qualifier, and that would also be okay. But while it makes recognizing an eyelike structure easier when speaking about eyes, it still muddies the metaphorical homological waters. Granted, saying that the scallop has over one hundred *mata*, or that a fruit fly has complex *oga*, or that a vertebrate *jicho* has a lens, is a lot harder to understand and process (unless you speak Swahili, Swedish, and Tongan) than saying that the scallop has over one hundred *eyes*, or a fruit fly has complex *eyes*, or a vertebrate *eye* has a lens. This sounds a lot like a word game and to a certain extent it is, but it is an important word game that gets us to realize how amazing nature really is when it comes to solving problems, especially with respect to light detection and color vision. Rather than saying nature solves a problem with the same solution, it is more precise, accurate, and impressive to realize that nature solves these problems with different solutions, all of them equally innovative and equally novel.

Eyes, No Eyes, Eyes, No Eyes, Eyes . . .

Light-sensing organs in animals are all predicated on the same basic molecular ground plan. In that sense they all start out with what biologist Sean Carroll has described as the same molecular or genetic toolbox—more or less. This toolbox includes several important and highly conserved genes that control developmental pathways in animals. In addition, part of the story of eyes is historical contingency. A famous experiment reported by Gavin de Beer, which is improperly cited by creationists as an argument against evolution, will serve to show how this contingency works. De Beer was a polymath and worked on a multitude of organisms and problems in developmental biology and zoology. He describes one important experiment using *Drosophila*, the tiny workhorses of experimental biology. The experiment started with flies without eyes, which were then allowed to evolve new "eyes." Getting eyeless flies is easy, as eyeless mutants in this lab species have been known since the early part of the 20th century. It is caused by a single gene called, with a lack of imagination, "eyeless." This gene was ultimately called a "master switch gene" in the 1990s by the late Walter Gehring, a molecular geneticist who worked on genomes in general and eye development in particular. De Beer did not have the luxury of 1990s molecular biology and knew only that the eyeless flies were the result of a mutation in that single gene we have been talking about (also known as "ey"). In the experiment de Beer described, the eyeless flies were allowed to breed, and eventually, after a few generations, flies with full-fledged eyes were obtained. The easy explanation for this result is that the eyeless gene reverted back to the wild type and hence the eyes returned. But further probing showed that the eyeless gene more than likely had little if anything to do with the flies getting back their eyes. The flies that regained eyes were mated to wild flies, and lo and behold, many of the offspring were eyeless. This could not be so if there was a mutation in the eyeless gene that reverted it back to wild type. De Beer reasoned that other genes were involved that were not the same genes that caused the original eyeless phenotype. While de Beer was more interested in that esoteric subject we mentioned, homology, this experiment introduced a conundrum. The eyes of the original line that gave rise to eyeless flies look exactly like the eyes that arose from the eyeless line, but they have a different underlying genetic and developmental background. Are they the same thing? De Beer answered, No. He reasoned that because the underlying genetic and developmental basis of the eyes before eyeless and after eyeless are different, they have to be different structures.

While this example is a bit convoluted and a bit unbelievable (which is why creationists tend to glom onto it as an antievolutionary argument), it is so because of the short period of time in which these changes occurred. De Beer was correct in stating that the eyes after eyeless and the eyes before eyeless are not technically the same structures, but because the time frame is so compressed, it is hard to swallow the story. However, if millions of years were interspersed between the loss of eyes and the regaining of "eyes," the story becomes a lot more palatable. Now, with all that said, there are still a lot of genes that make proteins that go into these fly eyes, expressed precisely the same in the flies with eyes before eyeless and the flies with eyes after eyeless. But the critical point is that the overall developmental process that generates the eyes in the two kinds of flies (eyes before eyeless and eyes after eyeless) is different. With this in mind let's take a look at eyes in animals and see why light detection and color detection in animals in general and us in particular evolved the way they did. Color vision will be prone to the same evolutionary conundrums. The reason we are talking so much about eyes here is that they are the structures that capture light that goes to the brain, where we and other organisms perceive color. It is important to trace the evolutionary history of eyes, because in many ways if you don't have eyes, you don't have color vision. As we will see shortly, color vision has a somewhat checkered history. Does this mean that color vision has evolved many independent times in the history of animals with eyes?

Cerebral Eyes

Some biological researchers focus on specific organisms, others on specific biochemical pathways, still others on entire ecological realms. Rarely do researchers focus on a term ripe for burning like homology, but that is what Detlev Arendt, a German evolutionary biologist, has done for most of his career. He was embroiled in a two-century-old biological argument as to whether the belly side of invertebrates is the same thing as the back side of vertebrates—the "invertebrates are inverted vertebrates" controversy (they don't necessarily have to be). His interest in homology has extended to eyes, and he uses the useful moniker "cerebral eyes" for eyes that are connected to a brain. When this qualifier is used, it clears up a lot of the clumsiness of *eyes* as a term, but it might be too accommodationist as far as clarity goes. It opens the door for plant, cyanobacteria, algal, and fungal eyes. Sometimes there comes a time when you just have to call a cerebral eye an eye and everything else something else. But do you?

The simplest cerebral eyes discovered so far are more than likely only two cells; one cell being a photoreceptor cell and the other what is called a "shading pigment cell." These cerebral eyes (they are connected to a neural ganglion or clump of neural cells) cannot tell an organism that there is a shape in front of it but can tell the organism where light is. Simply constructed organisms like zooplankton larvae of all kinds of marine invertebrates have the capacity to swim toward light, and it is these two-celled eyes that detect the light. Swimming toward light is an adaptation for these organisms, as they need the oxygen-rich water near the surface to grow, and knowing where it is is adaptive. Arendt and colleagues went a step further than just cataloguing the phenomenon and the kinds of cells involved. They showed a causal relationship connecting the two-celled "eye" with a neural response that is not really muscular but a motile reaction nonetheless. By manipulating the cells in this simple two-celled eye of the early stage of the marine worm species called *Platynereis dumerilii*, they showed that the phototactic cells directly impact the movement of the cilia of the plankton, which in turn mediates the movement of the animals toward light. The zooplankton of this species are not seeing a range of colors though. They might be seeing one color, as the pigment cells are detected by using a dye that is most efficient at 564 to 570 nm, or at wavelengths of cyan color, meaning that the pigment absorbs light of this wavelength most efficiently. Arendt makes the argument that these two-celled eyes are a remnant of the ancestral eye of a reconstructed organism called the "Urbilateria," or the mother of all bilaterian organisms. In other words, the common ancestor of all bilaterians (the urbilaterian) had these simple eyes made of two cells. This observation is a landmark discovery for animals and demonstrtates the first instance of a phototactic response mediating a well-understood and neurally connected reaction. What's more, these observations allow us to say that this more than likely happened in the mother of all bilaterians—The Urbilateria. If you still don't get it, think back to bacteria, algae, fungi, and those squishy animals at the base of the metazoan tree of life. In chapter 2 we said they all have phototaxis but no clear neurological processing of the phototactic response. The Urbilateria had the ur-eye, and it passed the ur-eye on to all of its descendants, and that means all bilaterian animals, including us.

To describe all twenty-five to forty different kinds of eyes that diverged from the urbilaterian mother of all eyes would be a daunting task and even more so to read about. But there are some highlights. For instance, within the extant arthropods (crustaceans, insects, spiders, and others), there are several basic kinds of eyes. The first is made of ommatidia with facets and is known as a "compound eye." Ommatidia are microphone-shaped structures with the facet at one end and the neural connection at the other. Light enters the facet, which serves as a lens, travels through the lens, and hits pigment cells, which react with the light. There are two kinds of cells in the rest of the ommatidia—optic and sensory. The first collects light and the second transmits the information (i.e., wavelength of the photons hitting it) and the second conveys the message to the brain. Hundreds if not thousands of these ommatidia make up a complex eye.

Perhaps the best example of a complex eye is in insects; visual systems like the one in the famous Gary Larson cartoon where a fly sees a woman raising a flyswatter in twenty hexagonal frames connected to each other. Larson is usually right on with his cartoons and animal anatomy, physiology, and biology, but in this case, he got the vision aspect of a fly wrong. Insects with compound eyes do not see thousands of images with their thousands of ommatidia; instead, their brains process all of the information from the ommatidia into a coalesced single image. Some insects have remarkably few facets in their eyes. Ants are notorious for having few facets. *Eciton burchelli* in particular has a single enlarged lens that resembles a spider eye. On the other hand, dragonflies have notoriously large numbers of ommatidia, up to thirty thousand in some species. Another kind of arthropod eye has a surface with spaced eyelets that are not in contact with each other like the ommatidia of insects. Centipedes and millipedes, along with the group of animals, Symphyla, that make up the larger group, Myriapoda, have this kind of eye. This arrangement constitutes a compound eye but quite different from the one we described for insects.

The third category of eyes has a transparent cuticle that provides inwardly directed lenses. These eyes are compound too but look smooth because the transparent cuticle covering the light-collecting cells in the various ommatidia acts like a corneal membrane. Horseshoe crabs have this third kind of eye, with around a thousand ommatidia. As you might guess, with a thick cornea and low number of ommatidia, they have pretty poor eyesight. The final kind of arthropod eye is typical of most chelicerates and is a single-lensed, or non-compound, eye. Spiders for the most part have eight of these non-compound eyes and see relatively poorly. All of these types of eyes force us to regard them as different and ask, Did they all come from the same ancestral kind of eye, say from the Ur-Arthropod? Nicholas Strausfeld and colleagues think they have solved the problem. They state that "there was an Ur-Arthropod eye that had the common ground plan of all arthropod descendants." Other animal "eyes" vary spectacularly. Molluscs (clams, oyster, octopus, and chitons) take the cake, with up to eleven different origins of eyes. Molluscan eyes range from eye cup, or pit eyes, with the photoreceptor cells exposed to the open, to compound eyes like hexapods, which we discussed above, to pinhole eyes, to eyes with mirrors, to closed lens eyes that are present in vertebrates like fish. If the structure of eyes tells us anything, it is that the problem of vision has been solved many times over in animals on this planet because detecting light is such an essential part of survival on this planet.

What about our human eyes? Researchers have determined that our eyes are part of a long evolutionary progression that started in the common ancestor of all vertebrates over 500 million years ago. And it is also clear that the vertebrate eye hasn't changed much since bony fish branched off the vertebrate tree of life. Unlike the invertebrates we examined above, vertebrate eyes appear to have found a common ground for how eyes should look and work. This doesn't mean that vertebrate eyes are boring though.

If you were to stick a pin directly into the center of your eye (please, do not try that), the tip of the needle would pass through the cornea, the iris writ large, the pupil, the lens, the vitreous body, the retina writ large, and the fovea. Ouch! That pin just travelled through all of the structures that light will travel through. The cornea is a protective sheet of tissue that covers the eye and has five discrete layers of tissue. Despite the five layers of tissue, light passes through the cornea with ease on to the iris. The iris is a relatively large layer of the eye in which the pupil lies at its center. The pupil is the muscle of the eye. It has two actions directed by two muscles. Smooth muscles dilate or expand the iris, and the pupillary sphincter muscle controls the size of the pupil. Both are important in regulating the amount of light that eventually reaches the retina. As we will see in chapter 4, the iris has loads of pigment in it, and this pigment controls eye color. Once the light passes through the pupil it is directed to the lens of the eye. The lens is biconvex, like two parentheses without words in between them: ( ). It is transparent, and, as its name implies, it focuses light coming into the eye onto the retina. Once the needle leaves the lens it penetrates the vitreous body. This part of the eye is a clear, gelatin-like structure that separates the lens from the retina. The retina is where all of the real action lies. It is a sheet of specific cells, which we will discuss shortly. These cells collect the information from light waves and send them on via the optic nerve to the brain. If we were a little off to the left in poking the needle into our right eye, we would have traversed through a structure called the "optic disc" and right into the optic nerve. But because our aim was true, we hit the retina. In particular we hit a field of cells packed much tighter than other areas of the retina called the "fovea." The fovea is where most of the light coming into your eyes is focused. The optic disc is important as it is the spot on the retina where the optic nerve is positioned, and this area is devoid of photoreceptor cells. Because there are no photoreceptors in this disc, no information from light comes from that part of the retina ever, and it produces what is called a "blind spot." All in all, the eye is a pretty amazing structure, but one that can be completely explained by the evolutionary process. To get to how the information collected by the retina gets to the brain, we will return to the optic nerve later in this chapter.

Who's Got What Opsin?

Based on what we know about bilaterians, in addition to the weird placozoan opsin we talked about in chapter 2, there are three major kinds of animal opsins: ciliary (c-opsin); rhabdomeric (r-opsin); and Go-opsins, which are a complex group of opsin-like proteins. To understand this diversity of proteins for detecting light within a specific kind of protein, we need to know a little about the structure and function of opsins. Opsins belong to a very large category of proteins that molecular biologists call G-protein-coupled receptors. In the evolution of eukaryotes on this planet, this kind of protein has been very successful in the face of natural selection, as they are the largest and most diverse family of proteins in eukaryotes. They are also known as "GPCRs." Let's break that name down.

We first need to know that a protein is simply a chain of amino acids coded for by genes in an organism's genome. Proteins also like to fold into three-dimensional structures. In so doing they form pockets and other places where smaller molecules can interact like puzzle pieces. For instance, the major protein in our blood is called "hemoglobin." There are several types of hemoglobin in your red blood cells, depending on the developmental stage you are in. Two each of two of these globin proteins, named alpha and beta, bind together to form a four subunit (also called "tetrameric") large globular protein, hemoglobin. Iron atoms, of which there are plenty in our bodies, bind to a small molecule called "heme," which sits in a pocket of the beta globin subunits in the hemoglobin molecule. Oxygen is held onto by heme and is effectively moved around the body through our circulatory systems, delivering this important energy source to other cells in our bodies. When oxygen reacts with the iron atoms bound up in the hemoglobin, it makes a red color. The important point is that if the beta globin protein is altered in shape because of, say, a change in the amino acid sequence as a result of mutation in the beta globin gene, the change in shape will prohibit oxygen from binding to the hemoglobin in the right place and in turn prohibit oxygen from being moved around the body. Individuals with this kind of mutation are anemic as a result of not moving enough oxygen throughout the body.

We have thousands of different proteins coursing through our cells at any given time (and even more if you count the billions of bacterial proteins produced by these interlopers on and in our bodies). The linear arrangement of amino acids in these proteins is what tells the proteins how to fold and where to sit in cells. As we've already seen, opsins have one end of the protein chain sticking outside the cell and the other end on the inside. The rest of the protein runs in and out of the membrane of a photoreceptor cell seven times. The chromophore, a small molecule called "retinal," sits in a pocket in the opsin molecule like a spring-loaded lever waiting to be tripped. Retinal is made up of twenty carbon atoms and a bunch more hydrogen and oxygen atoms. When it is inserted in the opsin, it looks like a bent bobby pin, as it is literally angled at the eleventh carbon. This bend is caused by a type of chemical bond with the other carbons, called a "cis bond." When a photon hits the opsin and is captured, it straightens the retinal out as if it had been a spring ready to be released. This straightening out pushes outward and changes the shape of the opsin molecule, which then makes the opsin more attractive for other reactions in the interior of the photoreceptor cell. Which brings us to the next part of GPCR, the actual G protein.

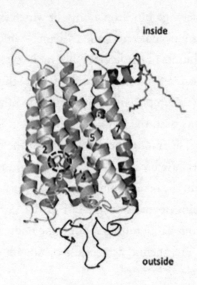

Figure 3.1. 3-D model of an opsin protein. Each of the seven transmembrane domains are numbered 1 thru 7. The inside direction of the cell and the outside direction of the cell are indicated in the figure. The dashed circle indicates where the chromophore is complexed with the opsin. *Drawing by Rob DeSalle.*

G proteins float around the inside of the cell and are activated by interaction with membrane-spanning proteins like opsins. The membrane-spanning proteins usually trigger a response on the G protein by changing shape. G proteins are tied up in a cellular response called "signal transduction." This term is broad reaching and can mean many things to researchers, but one thing for sure is that there is a response and a signal and this signal is essential to understanding color. These proteins are also known as "guanine nucleotide binding proteins," and they act as molecular switches. The question then becomes what do they switch on and off? Usually some stimulus on the outside of the cell like ionic concentration or light is the best answer to the question. They are called "G proteins" because they bind to and remove hydrogens from small molecules called "guanine triphosphate," or GTP. When the hydrogen is removed, the molecule is called "guanine diphosphate" (GDP). You can think of these three things—the G protein, GTP, and GDP—in the following way. When the protein is bound to GTP, the switch is in the on position, and when bound to GDP, the switch is in the off position. The difference is a simple hydrogen atom. The G protein is kind of misnamed, because sometimes it is not a single protein but rather a complex of proteins. Nonetheless, whether it is a single protein or a complex of proteins, GDP is bound to it in its "off" state.

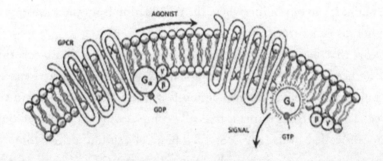

Figure 3.2. Diagrammatic representation of the GPCR complex. *Drawing by Pat Wynne, included with permission.*

How do G proteins switch from "on" to "off"? As we point out above, a receptor is coupled to the G protein in the nucleus of a photoreceptor cell. A retinal molecule sits snuggly in a pocket of the opsin photoreceptor protein (also known as a "GPCR"), where it is easily triggered by light. When light hits, retinal then flexes itself and changes the structure of the opsin, which in turn causes a reaction with the G protein it is coupled to. This G protein has GDP coupled to it and hence is in the "off" position. The reaction results in the conversion of GDP to GTP, which puts the G protein in its "on" state and releases one of its subunits into the inside of the cell. If enough of these reactions occur as a result of the external stimulus of light, then the concentration of the released subunit gets high enough to trigger other reactions in the cell. These end reactions signal nerve cells, which then transfer the information of the initial interaction of light and retinal in the nervous system. A signal is then sent on down the line, usually to the brain. G proteins are connected to a wide array of receptor proteins, not just light receptor opsins, and this way of turning things on and off as a result of some external stimulus has become a major theme in the way cells communicate with each other in complex eukaryotes. These chemical reactions are important because they carry signals from the retina of the eye to the brain. It's in our brains where the information from the wavelengths of photons gets turned into color.

Now that we know what the proteins involved in photoreception do, we need to get to the specific proteins to understand color vision. As we pointed out above, the real action is in the retina, especially in the fovea. The vertebrate retina is basically composed of two kinds of long, thin cells called "rods" and "cones." These cells sit like a thick forest of trees pointing outward from the retina. One type of cell has a rodlike outward-pointing tip, and the other a cone-shaped tip (it isn't hard to guess which is named what). They have different light-detection properties because of the kinds of opsins restricted to each. In addition to some basic differences in cell shape between rods and cones, their functions are quite different. The reason they function differently is that they have different kinds of opsins embedded in their membranes. There are four kinds of nonbacterial opsins—r-opsins, or rhabdomeric opsins; Cnidops, or cnidarian opsin; Go/RGR opsins; and the c-opsins, or ciliary opsins. While we have all of these opsins in our opsin repertoire (except for Cnidops, which is found exclusively on cnidarians), the one category that interests us the most for color vision is the c-opsin category, called such because these opsisns embed themselves into the cell membranes of the ciliary rod and cone cells. C-opsins are an invention of the urbilaterian, which arose over 500 million years ago, and they include the proteins we humans use for color vision. As we will see in chapter 4, these opsins have ended up in many different combinations in bilaterian animals, but the combination that our lineage settled on is one that gives our species a pretty broad range of color perception. The c-opsins include rhodopsin and the long, medium, and short-wave opsins. Remember those bacterial rhodopsins we talked about in chapter 2? They actually have the same name as the bilaterian rhodopsins, but in fact, their homology to eukaryotic rhodopsins is suspect. They have somewhat decent similarity in their shape, in addition to the seven membrane threading pattern. However, when the basic building blocks of bacterial rhodopsins and the four bilaterian opsins are examined, there is little, if any, similarity, indicating that they are not homologs of each other. Sound familiar? In this case, the rhodopsin proteins of bacteria and bilateria are analogs (as Darwin would put it), like the eyes of cnidarians and the eyes of mice.

There are five kinds of c-opsins that function in our retina. They are distributed among the rod and cone cells, but not evenly. Rod cells only have rhodopsin, while the cone cells have four kinds of photoreceptive opsin proteins sitting in their membranes. Because these opsins are specific for reactions with photons at specific wavelengths, they are named for the wavelengths they get most excited at. Another way of saying this is that the optimal wavelength is the one where the opsin is best absorbed. Using perhaps one of the simplest and most convenient protein naming systems around, these opsins are Long Wavelength (LWS), Medium Wavelength (MWS), and Short Wavelength (of which there are two—SWS1 and SWS2). Rhodopsin reacts with light at wavelengths of about 500 nm. These proteins are very sensitive and detect monochromatic hues. Their sensitivity is so great that with acclimation they allow us to see in very dim light. They work by light hitting the rod cell membrane in the retina and triggering the release of retinal from the protein. As we said, the intensity of light needed to induce this reaction is very low. As your eyes adjust to dim light, these very sensitive photoreceptor cells need to reset—in other words, get the retinal spring back in place so more rhodopsin reactions can be made. Have you ever had someone shine a flashlight in your face after you have acclimated to the dark? You lose your sight for a few seconds if not longer, and it takes you even more time to acclimate to the dark again. The reason for this is that rhodopsin is very sensitive to the bright light, and a large number of rhodopsin proteins release their retinal at the same time. Because it takes some time to reset the retinal back into the rhodopsin protein, there is a lag time before the rhodopsin proteins are adequately reset, which then allows you once again to see in the dark. While the rod cells' light repertoire is intense but somewhat limited, cone cells, on the other hand, are more versatile, though less sensitive to light intensity. Their versatility lies in the way they combine their signals to the brain to give us the color we see in nature.

The four other opsins are much less efficient at reacting to light than rhodopsins, which is why seeing colors in the dark is quite difficult. The four opsins (LWS, MWS, SWS1, and SWS2) reside in the cone cells of the retina. Because there are four opsins associated with different cone cells, each cell can react individually to four different ranges of wavelengths of light. These proteins are also known as retinal chromophores because they bind retinal and release it when hit by light of certain wavelengths. Note that we say "wavelengths" because any chromophore actually reacts to a range of wavelengths of light. So, for instance, the two SWS proteins activate most efficiently with light of wavelengths at 420 nm, but they will respond, albeit weakly, to wavelengths as long as 540 nm and as short as 360 nm. Likewise, for LMS opsins, their peak excitement comes when interacting with light of wavelengths between 400 nm and 680 nm, with a peak at 564 nm. Finally, the optimal wavelength for MWS opsins is 534 nm but also with a range of wavelengths from 400 nm and 650 nm. The range of wavelengths that our visual system can detect is therefore between about 360 nm and 680 nm. When rounded, that coincides with the widely cited 400- to 700-nm range seen in the literature.

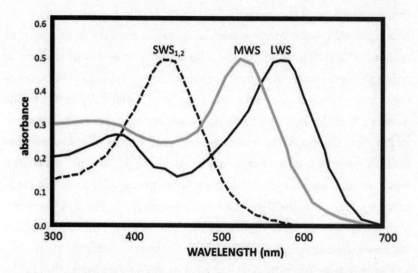

Figure 3.3. Absorbance and wavelength of SWS1, SWS2, MWS, and LWS opsins. The dashed line represents the absorbance of the two SWS opsins (blue). The gray line shows the absorbance of the MWS opsin (green), and the solid black line shows the absorbance of the LWS opsin (red). *Drawing by Rob DeSalle.*

The overlap of wavelengths that the four opsins cover ensures that information from light in this range is collected by the retina. But we have a difficult, if not impossible, time collecting information from light of wavelengths shorter than about 300 nm and longer than 700nm. Other organisms have evolved the capacity to detect light waves outside our 400- to 700-nm range, and we will discuss some of these in chapter 4. Within the 400- to 700-nm range, the overlap of wavelengths of the four opsins makes for some interesting signals that are sent to our brain for processing.

For the moment let's not worry about where the light hitting your retina comes from. In other words, don't worry about what you are looking at or what color it is; instead, let's concentrate on what happens to the light as it hits your retina. We have already discussed what happens when light hits the membrane of the rod and cone cells and impacts the opsin protein sitting in the membrane. That was with just a single opsin protein. Imagine the photons coming into your eye and being focused on the retina by your cornea and lens. These photons will hit millions of opsins in millions of rods and cones. Remember too that there are multiple kinds of cones in the retina, which will have opsins that respond to the light depending on the wavelength of the light hitting them. Finally note that the ranges of wavelengths that the opsins respond to represent probabilities. When we say that the peak wavelength of an opsin is 564 nm and it ranges from 400 to 680 nm, we are making a probabilistic statement about the capacity for the opsin to capture a photon. If the light has some wavelengths at 564 nm, there is a high probability the LMS opsin will react to it. It's not a probability of 1.0 but nonetheless a high probability. Likewise, as the wavelengths get further and further away from the peak wavelength, the probabilities fall. On the shorter side, wavelengths of 400 nm have very low probabilities, but these are not zero probabilities. Shorter than 400 nm, the probability is so low that it is effectively zero.

Each cone cell produces an output of signals. Taken together for the whole retina, that becomes an immense amount of information. The average human retina has about six to seven million cone cells, with most of those concentrated in the fovea, that little area of cone cells in the center of the retina. There are about ninety million rod cells, and these are also interacting with photons and delivering signals into our visual system. How is all of this information in the form of neural impulses being processed by our visual system? That question requires us to understand two important words in the visual system vocabulary—*univariance* and *opponency*.

The opsins in your retinal cells are going to respond to light wavelengths uniformly every time one captures a photon. The probability of a photon being captured changes with its wavelength for any given opsin. It isn't the actual wavelength of the photon, then, that we are "seeing" or that our visual system is processing, but rather the number of photons captured by the retina's cone cell opsins. This phenomenon is called "equivalency" and was first articulated by D. Ewart Mitchell and William Albert Hugh Rushton in 1971. This principle tells us that our cone cells on their own are color neutral. The single cells and their opsins don't care what wavelength the photons are—they respond in a set way, which is dictated by probability. Neither does our visual processing system care, because all it is receiving is output from the cone cells. Rather, comparing the number and kinds of cone cells responding to light is the first step in understanding color vision. Opponency is the remaining concept to delve into.

Scientists like to fight with each other almost as much as they need to collaborate. Both intellectual scuffles and cooperation are essential for science to progress, according to some historians of science. Such is the story behind the opponency theory of color from mid-/late-19th-century Germany. The apposition of scientific intellectual schools in Germany during this period is well known to have advanced science in many areas, but color vision was particularly impacted by the acrimonious interactions of two scientists who battled with each other over the nature of color vision. The two scientists involved were Hermann von Helmholtz and Ewald Hering. These two men of very different backgrounds battled bitterly over how color vision works. As with many scientific scuffles, both were correct in their stances but stubbornly, and sometimes insultingly, insisted the other was wrong and literally off their rockers. Helmholtz was of an upper-class upbringing and somewhat of a prodigy in scientific circles of the time, while Hering had a more difficult early career that established him as an expert in vision research and psychology. To say that they didn't like each other would be an understatement. They disagreed with each other on many aspects of vision, and in one particularly acrimonious exchange about eye movements, Hering sarcastically suggested that one of Helmholtz's theses about eye movement was the result of Helmholtz not realizing he had fallen asleep and in so doing had rotated his head. Helmholtz responded by admitting Hering was annoying but dug deeper into Hering's personality with this insult:

"Mr. Hering has annoyed me considerably with his impertinent ways of judging other people's work which, in part, he has not taken the trouble to understand properly. . . . He has been, as I have heard, mentally ill and this has until now held me back from bringing him down, which he has at times deserved."

While this feud is not usually mentioned in the top ten scientific fights, the scuffle was one of the more acrimonious in terms of personalities and the dogged persistence of Hering, who felt he had an axe to grind when Helmholtz heightened the conflict. Even after they both died, their students and colleagues carried the debate on for a while. So, what was the fuss about?

Some background is needed, and it begins with the work of Thomas Young, who proposed that all colors are based on three primary ones—red, green, and blue. This very simple but absolutely correct hypothesis about colors was part of the dispute between Hering and Helmholtz. Next, James Clerk Maxwell, the great physicist, demonstrated that any color can be generated by mixing the three primary ones, as hypothesized by Young. Both of these ideas were incorporated by Helmholtz into his thesis about color vision, which stated that the human eye should have three kinds of color receptors, and that shades of black and white also perceived by the eye would enhance or reflect the amount of light coming into the eye. Hering, on the other hand, suggested that the color perception system of humans was based on his idea of opponency. Opponency refers to the idea that colors are perceived by three couplets of opponent colors—red/green, blue/yellow, and white/black. Whereas Helmholtz used Maxwell's empirical work on primary color mixing, Hering really didn't have an empirical leg to stand on. The two men's theses sound a little different, but they are both correct, as we now know that indeed there are three color receptor types in the retina (LWS, MWS, and the two SWS receptors), and indeed these are combined into three opponent comparisons that the eye makes in sending the information off to the brain. Both Hering and Helmholtz were correct. Their argument was sort of an apples and oranges one, which is usually the case when both parties of an argument are correct.

It is important to note that the retina isn't just made up of rod and cone cells but actually has at least three layers, each of which has a specific job in processing light entering the eye. We already have seen how the three receptors work, and that it is the role of the cone cells to gather this information. Rod cells, as we have seen, are distributed throughout the retina (mainly missing in the fovea). These rod and cone cells make up the first layer of the retina. The opponency—that is, the comparison of the amount of signal the cone cells transmit during color vision—of these primary color receptors and their signals occurs outside the cone cells, in the second and third layers of the retina behind the rod and cone cells. The second layer is made up of bipolar cells. These cells are the first to receive the signals from the cone cells, and they make direct comparisons of signals coming from the different cone cells, based on opponency. This information is transmitted to the third layer, made up of ganglion cells. The comparison of the cone signals produces what neurobiologists call "opponent channels" and is the basis of how opponency works.

There are three kinds of opponent channels. The first two concern color opponency, and the third is a dark/light opponency, as Hering hypothesized. The first opponent channel is, as Hering suggested, a comparison of signal from red and green receptors, or, as we now know, between LWS and MWS receptors. The direction and magnitude of the opponency is evaluated in this channel in the bipolar cell layer. The other color-based opponent channel as Hering hypothesized also exists, with one twist. Since there is no "yellow" receptor (no yellow opsin at all), the opponency is evaluated between the signal from the short wavelength receptor (SWS) and the sum of the long and medium receptors (LWS + MWS). Once the opponency of these three channels is assessed, the proper signals that represent the opponency are transmitted to the third layer of the retina made up ganglion cells. The ganglion cells' axons then direct the signal out of the retina into the optic nerve, which is a large bundle of neurons leading from the back of the eye to the lateral geniculate nucleus (LGN) in a part of the brain called the "thalamus." In the LGN the color channel and light/dark channel signals are combined with other elements of the objects being viewed by the eye and sent on to the visual cortex, a large collection of neurons toward the rear of the brain. The LGN and visual cortex process the information further, and this results in perception and recognition, where our memories and other higher brain functions help interpret what is being seen by the eyes.

The visual cortex of the brain has been studied by scientists for well over a century. Much is known about how the visual cortex is structured and where the electrochemical impulses or action potential from other parts of the body end up and how they interact to produce perception. We know most of what we do about vision in general and color vision in particular from experimentation using three major "technologies." The first consists of clever psychology experiments done on humans. The second involves macaque monkeys and the ease with which their brains can be explored during visual perception. This second approach is akin in many ways to studying brain lesions in humans; the difference being human brain lesions are sad happenstances, while the study of macaques can be done directly and intentionally. The final approach involves the imaging of human brains using magnetic resonance imaging while a subject performs some task or is exposed to some stimulus (functional, or fMRI). Charles E. Connor, a neurobiologist who studies vision, suggests that these three approaches are the *what* (psychological experiments); the *where* and *when* (anatomy studies and imaging using MRI); and the *how* (animal experiments focused on brain physiology, anatomy, and neural connections and the analysis of human brain lesions) of vision.

The what

Psychology and cinema theory are probably the two college majors most parents dread hearing the most ("What kind of job are you going to get with a degree in psychology!?"). But psychologists who study vision and other brain functions have developed some of the most creative and well-designed experiments in science. The typical psychology vision experiment often involves some optical illusion or optical phenomena and a way for the experimenter to assess the response of subjects to the illusion. As Connor says, psychological studies address what the brain's visual limitations are "by measuring the perceptual and cognitive capacities of human subjects." Here are some "whats" that have been learned from studies of psychology and vision.

If you go to the kitchen, pull out an empty Jell-O mold, and set it down so you are looking straight on to it, sooner or later the indent in the mold looks like it is sticking out. This is an especially striking illusion when the Jell-O mold is a human face. We invert the figure from being indented to sticking out because we are used to seeing faces and other forms in three dimensions, which we then fill in. Seeing a face or another form that caves in doesn't make much sense, so we interpret what we are seeing as a three-dimensional convex figure. This experiment prompted psychologist Richard Gregory to suggest that vision is a top-down process. In other words, when we receive information from the eyes, we go immediately to a big-picture interpretation of the phenomena tickling our senses in our brains. Gregory suggested vision as a top-down process because we lose much of the information collected by our sense organs on the way to the brain. In this case the eye collects the info, processes it, and sends it to the brain. However, along the way, much of the information gets lost or is not transmitted, so there is some calculation or interpolation our brains do from a big-picture perspective to interpret what we are seeing. Why? The simple adaptationist answer, which makes some sense in this case, is that we do this in order to make quick decisions about what we are seeing. Our brain makes the best estimate of what it might be that we see, and this can best and more quickly be accomplished with an overall top-down context. Another illusion that Gregory felt shored up the top-down idea is the Necker cube. Everyone has seen this illusion, where a drawing of a cube can flip from being in one orientation to another. According to Gregory, this happens because both orientations are equally plausible to our brains, so we can easily flip from one to the other. This illusion is unlike the Jell-O mold, where our brains make the calculation and overwhelmingly decide "that object is sticking out toward me."

Of course, if there is the potential for top-down processing, this means that there also might be a bottom-up way of doing things. According to this approach, perception results from taking the component elements of what the visual system collects. The brain then constructs a story that is taken as perception by combining and evaluating the construct. This latter idea was the brainchild of James Gibson and has equally compelling reasons for why it might be the way vision works. Gibson suggests that there is no interpolation or quick decision made by the brain to perceive something a certain way. Rather, there is a calculation, but it is a straightforward one, like adding all of the information up and coming to a conclusion based on the summation of information. Bottom-up processing doesn't rely on filtering the information like top-down does. Bottom-up processing has oftentimes been called the "ecological approach" to processing because it incorporates the environment and other aspects of vision that top-down approaches don't.

You get what you put into it with bottom-up processing. It is an appealing alternative for three reasons, all of them evolutionary. First, the illusions that Gregory cites as support for top-down processing aren't part of our real world; they are made-up phenomena that our visual systems really never had to face, so there is no selection pressure for our brains to have responded to with top-down processing. If there are no illusions like the Necker cube or the Jell-O mold in nature, there is no need for top-down processing to evolve. Or at the very least, there aren't adequate data to test whether top-down processing is valid.

Second, Gregory's top-down processing is sort of an evolutionary just-so story. Such stories are not always the best way to interpret what we see in nature. The just-so story for visual processing goes like this. We have top-down processing because it afforded our ancestors who first accomplished it a selective advantage for survival. Quick decisions when you view something can be the difference between life and death when nature is red in tooth and claw. Just-so stories really need to be avoided when thinking about these kinds of traits in organisms, because the trait could just as easily have arisen as a side product of selection on other traits or even randomly.

Finally, it doesn't make much sense in the visual system to have such a precise mechanism as the retina, opsins, rods and cones, channel opponency, and other neurophysiological processes, and then ignore 90 percent of it just to make an estimation of what we think we are seeing. Our brains, as our colleague Gary Marcus suggests, are kluges: messy, clunky, with too many moving parts, but they work (for the most part). Evolution really hasn't streamlined much of our brains, so why streamline how this high-quality information is processed, when bottom-up processing can do the trick in a more parsimonious fashion?

The "whats" here are relevant to color vision too, because equally compelling experiments with illusions and other psychological approaches have been used to tease apart color processing in our brains. More than likely, as with Hering and Helmholtz almost a century earlier, the real process is a combination of the two (top-down and bottom-up) processes. Gregory and Gibson carried out their conflict in about the most non-acrimonious of possible interactions. In his writing, Gregory always gave Gibson his due, and it is said that Gibson was less interested in winning the argument than he was in getting researchers to think about the question harder and more clearly.

The where and when

Connor's second kind of approach to understanding color vision establishes the neuroanatomy and physiology of the brain. As we pointed out above, the visual information is ultimately sent to the visual cortex in the rearward occipital lobe of the brain. This region has been extensively mapped in humans but even more intricately so in rhesus macaques. And here homology rears its head once more. If we are trying to understand the function of our brains by comparison to a macaque brain, metaphors are not of assistance. In order to transfer our knowledge of macaque brains to something meaningful about our own, the homology of the brain regions needs to be established. We can learn some things from nonhomologous structures and physiologies, but the more straightforward way to impute function is for homology to be evident.

Fortunately, macaques and our species are closely enough related that homology of brain regions can very easily be determined. Macaques also have the same opsin system we do, with one difference. Most humans are trichromats—we have three visual opsins. A good proportion of macaques are trichromats too, but there are a considerable number in any given population that are dichromats (having only two of the visual pigments). The astute reader might ask, Why not use chimpanzees—after all, they are our closest living relative? Using chimpanzees is much more ethically fraught than macaques, so this Asian monkey has been the target of much of the "where and when" research.

Human anatomy and physiological experiments on lower primates like macaques have given researchers a pretty clear picture of the neural pathways in the brain that process visual stimuli from the environment, which are collected by the retina. The visual cortex has five areas that are demarcated because of their locations and their functions. These areas are called "V1 to V5." Of these areas, V1 is the first stop for visual and color information in the visual cortex receiving impulses directly from the LGN. The information then proceeds to the V2 area, which then sends the info along to V3, V4, and V5 in different pathways, which we discuss below. The information is then sent out to at least twenty other locations in the brain for more processing. You might think it difficult to make sense of all this wiring, but neurobiologists have delineated three major and parallel pathways that process visual information.

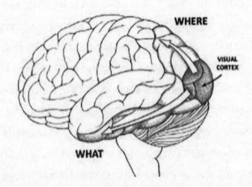

Figure 3.4. The visual "what" and "where" neural pathways in the human brain and the visual cortex. *Drawing by Pat Wynne included with permission.*

The first pathway courses through a cell layer in the LGN in the thymus, connects to V1, then projects on to V2. This pathway, known as the "parvo-cellular pathway," is made up of what neurobiologists call feature detectors, making this pathway particularly well suited for detecting fine detail. The second visual pathway is where neurons in V1 connect to V2 once again, but further projections are made out to V3 and V5, where shape and movement, respectively, are detected. We discuss the third pathway below, as it has been thought to be involved in color vision.

The how

Now that we have the what, when, and where at hand, we can easily go to the "how," right? That, however, would be asking too much, because putting this all together is one of the more difficult problems in neuroscience and—as we will see in chapter 8—philosophy. The problem of putting all of this information together to give a coherent single image that we "see" is called the "binding problem." Binding is the process whereby our brains connect colors and shapes our eyes tell us are out there with specific objects that are out there. There are two sides to binding. The first is called "segregation." It is the process whereby we assign characteristics of objects to the correct objects. Say we see two objects of different shapes, and our eyes deliver information to our brains that the shapes are blue and red; segregation is the process whereby we assign the red color to one shape and the blue color to the other. Perhaps if we only saw one object at a time we wouldn't have to segregate this information and the binding problem would be a little simpler. However, our environments are full of objects in need of identification. And so as more and more objects are in our field of view, the so-called binding gets more and more complex. Another aspect of binding is called "combination." This aspect of binding is how all of the input is put together; this includes the color, the shape, the texture, the background, aspects of the object in memory, and the emotional context of the object. The binding problem is indeed not just a little problem but a big one, and it has been discussed and debated for centuries (without calling it binding) by philosophers and neuroscientists. We will return to this problem in the context of color in chapter 8.

Considerable experimental progress has been made on the binding problem or, in other words, on the "how" of vision. How do we know how color vision works? Again, we go back to macaques, where researchers can pinpoint single neurons or areas of the brain using surgical techniques. They measure the neural impulse—what neurobiologists call action potential—that runs across these isolated cells while the monkey is exposed to some environmental stimulus. This very powerful approach has been used to show how the V1–V5 regions of the brain are connected to each other in vision. Color processing from the retina to the LGN and onto where information enters the visual processing pathways in the back of the brain is well explained but complex. It involves the relay of the initial three-color, three-channel information. Once the somewhat fragmented signals from the retina enter the V1–V5 pathways in the primary visual cortex, things get weird and harder to explain.

The complications start in V1, which is a junction point for the color information coming into this part of the brain. There are several neurons in the V1 tract that respond better to signals emanating from particular wavelengths of light than others. This phenomenon is referred to as "color tuning." However, these color tuning responses are not consistent if the environment is not consistent. For instance, some of the cells will respond better if bright light is the background for the color information. If the background light is low, then these cells might respond to a wider range of wavelengths of light. There are cells that respond in a consistent way in this region of the brain. These are called "double opponent cells" because they can keep track of light wave information in a very localized fashion. These cells occur in blobs (that is their technical name too) of V1 cells, and they compare the number of different photons coming from a scene that a person is observing.

The V1 blobs transfer the information that they have extracted from the input action potentials (those electrical signals that come all the way from the retina) to the V2 region of the visual cortex. Some of the cells in the V2 are also color tuned; they respond to color information, further refining the information until it is passed on to V4 and adjacent regions (at one time, V4 was thought to be uniquely dedicated to color processing). Because researchers now know that V4 also processes information about the form of what is being looked at, the area cannot be exclusively devoted to color perception. To complete the circuit, cells in the cortex of the temporal lobe are connected to V2. Some researchers think that the color information is then integrated with the spatial, shape, and form information to give a visual image to the observer. But how this is accomplished involves some handwaving.

It is important to realize that all of the pathways involved are complex, but the general flow of information from retina to LGN to V1 to V2 to V4 and finally to the temporal lobe is well established and integral to any understanding of how color is integrated into our overall image of what is around us. Stepping back and looking at the whole process, it is incredibly interesting to keep in mind that this pathway is populated by nerve impulses that originate with photons hitting the retina and a simple process of photon capture by a protein. This triggers a cascade of electrical impulses (action potentials) that become the currency of information in the pathways. All of the intermediate stops in the pathways are there to sort the information, amplify it, use it to compute comparisons, and eventually merge it with information about shapes and size and perspective to give a coherent image interpretable by our brains, which we then can use to navigate the environment around us. A truly remarkable system! It is a "klugey" system but one that works quite well. And a system that we can interpret as a product of the evolutionary process, as we will discuss in the next chapter.

4

The Colors of Evolution

The many colors in nature are astonishing. As we will see in this chapter, color is intertwined with the evolutionary process so intricately that one could teach an entire evolution course based on examples of color in nature. The study of color has been a key to the discovery and testing of evolutionary principles. In parallel, evolution has led to a wide color palette among organisms. How color varies and how it is produced in nature has been molded by all of the internal processes of evolution, including natural selection and genetic drift. There are many evolutionarily interesting phenomena in nature like aposematism, mimicry, camouflage, secondary sexual characteristics, and natural coloration phenomena like bioluminescence, pearlescence, iridescence, and fluorescence found in nature. All of these topics are great examples of the evolutionary process at work. In chapter 1 we looked at natural selection as a process driving evolution. Now we need to explain how the process works in principle with color.

Before Darwin, natural historians like his grandfather Erasmus Darwin, the famous French naturalist Jean-Baptiste Lamarck, and others knew something was going on in nature. That something was evolution. These natural historians saw patterns in nature that convinced them that nature was neither random nor created. In fact, the patterns suggested something more important and oftentimes directed. It was even called evolution well before Darwin started his voyage on the *Beagle* in the 1830s. Before Darwin, Lamarck came up with a process for how evolution might work; he was also the first to suggest an intimate involvement of the environment in the process. Today, Lamarck's ideas are touted as valid by those researchers who study how environmental factors can influence the genome (epigenetics). A close reading of Lamarck suggests that he had a prescient hold on how the environment influenced characteristics of organisms, but his ideas stopped far short of detailing an explanation for the evolutionary process.

Wallace and Darwin

The biggest step to understanding what was driving evolution was of course made by Darwin and another natural historian of the time, Alfred Russel Wallace. Both scientists used their sharp intellects and superior skills at observation to come to the same conclusion.

Both Darwin and Wallace travelled as young men to many exotic places. Darwin was included as a gentleman companion and natural historian on the second voyage of the HMS *Beagle* in its circumnavigation of the globe from 1831 to 1835. Wallace's journeys occurred during two periods of his early life. One trip to South America (1848–1852) and a second to the East Indies (1854–1862), mostly in the Malay archipelago, resulted in enormous collections of insects and vertebrates, some of which were lost when one of the ships he was sailing on caught fire. Both men were ardent naturalists who collected thousands of specimens and considerably increased the knowledge of biodiversity on the planet. Darwin named about 60 species (all of them barnacles), and Wallace named 307 species (12 palms, 120 butterflies, 70 beetles, and 105 birds). But what they had to say about color is most illuminating.

The controversy of who was first with natural selection doesn't matter much. Their seminal papers published in 1858 actually had a disclaimer at the beginning outlining the history of the idea of natural selection. It attributed the first kernels of the idea to Darwin's notebooks, which he had kept for twenty years from the 1830s into the 1850s, scribbling notes about natural selection more than twenty years before the publication of *On the Origin of Species*. We imagine that Darwin understood the primacy of publication. He had danced around the idea in his scientific papers to that point and had not specifically published the idea in a peer-reviewed journal (his diaries and notebooks were not peer reviewed or published until well after *Origin* had been published), but he simply played by the rules and simultaneously published his ideas with Wallace in the *Proceedings of the Linnean Society*. Darwin cared that both he and Wallace got credit, and he did not claim primacy. So, if it didn't really matter to Darwin, then we suppose it shouldn't matter to the rest of us and Wallace should get his due recognition as the coauthor of one of the most influential ideas of the past three hundred years. There is no doubt, though, that the ideas, writings, and details these two natural historians developed together are integral for understanding the process of evolution.

Wallace

There are copious references to color in Wallace's books and scientific papers. So, let's look at what these two giants of evolutionary biology had to say about color in their seminal works.

For example, on camouflage, Wallace noted in his studies of tropical birds: "Their plumage is so near the color of the foliage, that it is sometimes impossible to see them, though you may have watched a whole flock enter a tree, and can hear them twittering overhead, when, after gazing until your patience is exhausted, they will suddenly fly off with a scream of triumph." Wallace recognized that there was a connection of this organismal strategy for survival to natural selection. Color biologist Tim Caro has examined Wallace's writing in great detail and has identified six major categories that Wallace described as "functional questions about coloration that still demand investigation." These six categories are (1) protective colors, (2) warning colors, (3) mimicry, (4) sexual colors, (5) "typical colors," and (6) attractive colors in flowers and fruits. Of his six categories, protective coloration, aposematism, mimicry, and sexual colors have remained major focal points of modern research. Typical animal and some plant colors can be subsumed into the other four categories.

These categories of coloration in nature are indeed the basis of intense research even today and areas that we will delve into later in this chapter. But why make such a big deal out of Wallace constructing these categories? The answer is that Wallace invoked natural selection to delineate these categories. This demonstrates not only his hold on the natural process of selection but also that natural selection was real to him and an overarching way to understand diversity in nature. Actually, these categories have stood the test of time because Wallace based them on natural selection. But he wasn't just interested in categorizing things in nature. He, like Darwin, was looking for universals.

Consider Wallace's interest in color in the tropics. He noted that organisms were more brilliant in the tropics than in the temperate areas where he came from. This must have been a very striking visual change for him and others who came from Europe. He correctly linked color differences in the tropics and temperate zones to natural selection. In the following quote he sets the observation clearly: "The idea that nature exhibits gay colors in the tropics and that the general aspect of nature is there more bright and varied in hue than with us, has even been made the foundation of theories of art." Here he breaks from his natural historian focus and suggests that this observation has somehow been infused into how human art is created.

The intricacy with which he interpolated natural selection into his observations can be seen, where he compares the frequency of color patterns in the tropics against temperate climes:

> . . . In the tropics, a greater proportion of the surface is covered either with dense forests or barren deserts, neither of which exhibit many flowers. Social plants are less common in the tropics, and thus masses of color are less frequently produced. Individual objects may be more brilliant and striking, but the general effect will not be so great, as that of a small number of less conspicuous plants, grouped together in masses of various colors, so strikingly displayed in the meadows and groves of the temperate regions. . . .

Wallace invoked a difference of ecological conditions that molded the patterns of colored flowers in the tropics. His observation was that large stands of colored flowers are rare in the tropics but aplenty in temperate regions like the meadows of England. Instead, plants in the tropics need brighter, more efficient colors for survival because they are not "social" (here meaning "found in large stands"). What is remarkable about these ideas on color is his attempt at—and success in—synthesizing the observations with a mechanism. Both Wallace and Darwin succeeded in finding one of the processes that drives evolution because they were able to synthesize, to put things together from different levels of organization and biological complexity. Color had a lot to do with it.

Darwin

Darwin's reference to color was also voluminous. Like Wallace, he recognized many of the same major categories of the role of color in nature. For instance, he had this to say about the cuttlefish, one of the more versatile animals on the planet with respect to color:

. . . This cuttle-fish displayed its chameleon-like power both during the act of swimming and whilst remaining stationary at the bottom. I was much amused by the various arts to escape detection used by one individual, which seemed fully aware that I was watching it. Remaining for a time motionless, it would then stealthily advance an inch or two, like a cat after a mouse; sometimes changing its colour . . .

Many organisms use color as a means to blend into their surroundings. Most of them do this with permanent colors that allow them to do the blending in, as Wallace described with the tropical birds above. But Darwin mentions two organisms that have the remarkable capacity to change colors. How this amazing natural feat is accomplished will be discussed later in this chapter, but the mere fact that Darwin and Wallace recognized color as such an important cog in the machine of evolution is remarkable.

Darwin strove for the most precise way he could think of to characterize it in nature. Consequently, he took along with him into the field one of the first color dictionaries in existence, *Werner's Nomenclature of Colours*. This book was first produced in 1814 and was constructed like those paint charts in hardware stores showing a wide range of colors. For instance, the guide's author, Abraham Gottlob Werner, maintained that there were eleven shades of blue, ten or so shades of red, and even eight shades of white. The guide was expanded upon by Patrick Syme, an artist who added many more colors to it. The colors were poetically described; along with the description, each color was given a name and assigned a number. Finally, an example of an animal, vegetable, and mineral with that color was listed in tabular format. In the early 1800s, color printing was in its infancy, using an approach called "chromolithography." This approach used printing plates that could be doused with specific colors of ink and then printed onto paper or other surfaces. However, this was a poor facsimile for what a person viewed in nature, as only a few colors could be used in the chromolithograph. Like the monks who copied texts in the

Middle Ages, Werner had each color hand-painted in each copy of the book into little boxes that then appeared in the tables.

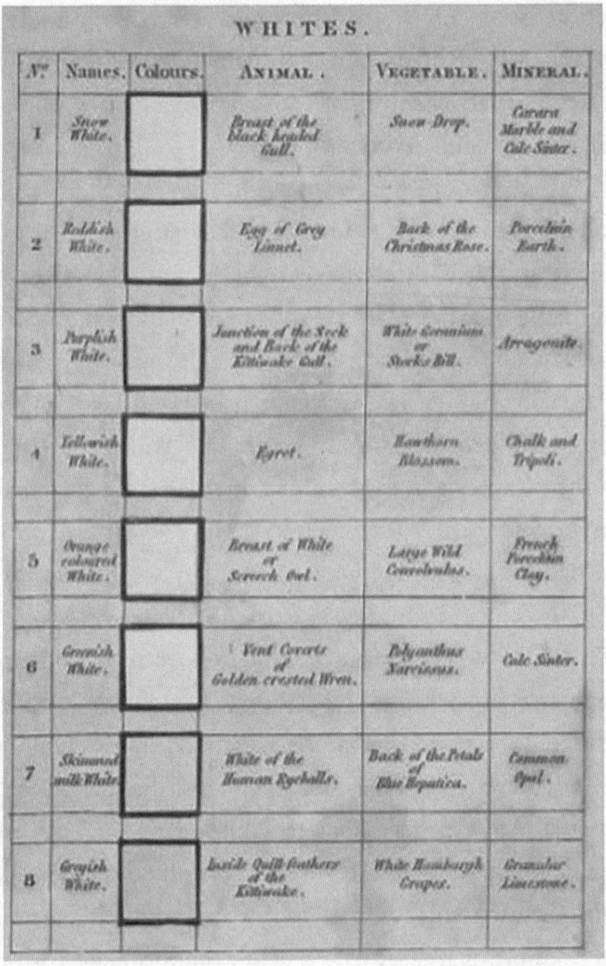

N°	Names.	Colours.	ANIMAL.	VEGETABLE.	MINERAL.
1	Snow White.		Breast of the black headed Gull.	Snow-Drop.	Carara Marble and Cale Sinter.
2	Reddish White.		Egg of Grey Linnet.	Back of the Christmas Rose.	Porcelain Earth.
3	Purplish White.		Junction of the Neck and Back of the Kittiwake Gull.	White Geranium or Stocks Bill.	Arragonite.
4	Yellowish White.		Egret.	Hawthorn Blossom.	Chalk and Tripoli.
5	Orange coloured White.		Breast of White or Screech Owl.	Large Wild Convolvulus.	French Porcelain Clay.
6	Greenish White.		Vent Coverts of Golden-crested Wren.	Polyanthus Narcissus.	Cale Sinter.
7	Skimmed milk White.		White of the Human Eyeballs.	Back of the Petals of Blue Hepatica.	Common Opal.
8	Greyish White.		Inside Quill feathers of the Kittiwake.	White Hamburgh Grapes.	Granular Limestone.

Figure 4.1. Page from *Werner's Nomenclature of Colours* for hues of white. Each of the shades has a column for its number (No.), the name of the color, a hand-painted version of the color, an example of the color in an animal, an example in a vegetable, and a mineral example. So for instance, the first color is catalogued as #1. It is known as "snow white," the animal example is the breast of the black-headed gull, the plant example is the snowdrop, and the mineral examples are Carrara marble and cale sinter. *Public domain.*

Today we have pantone systems such as the Pantone matching system. This system has 1,114 different colored reference tiles or spots that artists, designers, and other color oriented professions can use to tease apart hues in their work. This modern resource is very efficient, but for Darwin Werner's system was perfectly adequate for the precision he needed. Darwin felt he needed this resource to most precisely describe what he was seeing in nature. And indeed, even a cursory reading of his works reveals an amazing attention to color detail, sometimes beautifully precise, a precision that exists in everything that he wrote about. As an example, take his chapter on domestication in *Origin of Species*, where he used several pages to describe the various breeds of pigeons, a species that he bred and studied extensively starting in 1856. The extent of description of pigeons in this part of *Origin* is a whopping three thousand words long. Contrast this with the Wikipedia entry for domestic pigeons at a little over five hundred words. Color was a very important part of his descriptions for domesticated animals, and color almost led him to understand inheritance, something he never really got quite right. In the following from *Origin of Species*, he describes crosses of pigeons and uses color to attempt to decipher what he was seeing:

> . . . when two birds belonging to two distinct breeds are crossed, neither of which is blue or has any of the above-specified marks, the mongrel offspring are very apt suddenly to acquire these characters; for instance, I crossed some uniformly white fantails with some uniformly black barbs, and they produced mottled brown and black birds; these I again crossed together, and one grandchild of the pure white fantail and pure black barb was of as beautiful a blue colour, with the white rump, double black wing-bar, and barred and white-edged tail-feathers, as any wild rock-pigeon!

No wonder Darwin followed those observations with an exclamation point. Note that he is amazingly close to discovering recessive and dominant traits simply by following color. He attributed these observations to a concept around at the time called "reversion to ancestral characters," or atavism, but the patterns he observed could just as easily have been attributed to the more complex but satisfying process of inheritance with recessive genetic elements.

That color was susceptible to natural selection was a major theme of his writing on color. Darwin thought that because color was correlated with traits involved in survival, color itself would be subject to natural selection. He articulates this reasoning and gives us yet another factor involved in the action of natural selection:

> . . . and there is reason to believe that constitution and colour are correlated. A good observer, also, states that in cattle suscep-tibility to the attacks of flies is correlated with colour, as is the liability to be poisoned by certain plants; so that colour would be thus subjected to the action of natural selection.

Darwin also used the color of domesticated cattle and a thought experi-ment to dig down into how natural selection might work. He proposed the color of cattle left alone for generations as an example of natural selection in action. He describes a population of cattle that are of three different colors and then states, "One colour would in all probability ultimately prevail over the others, if the herds were left undisturbed for the next several centuries." Here, Darwin articulates that existing variation can slowly but completely be molded by natural selection.

His use of color from two of his books, *On the Origin of Species* and *The Voyage of the Beagle*, demonstrates that he was fixated on color as a major way to explore natural selection in nature. As we stated above for Wallace, in many ways Darwin also opened the door for almost all evolutionary study on color that we do today. When Darwin saw color, he also saw natural selection. In addition, he saw evolution as a direct contradiction of creationism. He makes the incredibly logical argument about creationism versus evolution based on color variation in plants: "Why, for instance, should the colour of a flower be more likely to vary in any one species of a genus, if the other species, supposed to have been created independently, have differently coloured flowers, than if all the species of the genus have the same coloured flowers?" The wonderful logic of this statement is a great argument against creationism, a battle that Darwin fought both personally and in print.

Color, natural selection, and populations

Color was one of the major natural phenomena that prompted Wallace and Darwin to gravitate toward evolution and natural selection as an explanation for the grand variation observed in nature. But in order to make their ideas timeless, the two naturalists needed to do more than just convince their audience of the existence of evolution. They also needed to pin down the mechanisms involved. One of the more obvious aspects of color in nature is its variation—not just among species and genera and vastly different kinds of organisms but, more importantly, variation within populations of organisms. The amazing variation of color in populations that these two naturalists documented is an obvious indicator that variation is a requirement for natural selection to proceed. Variation is essential for evolution of color and color-producing mechanisms to occur. Darwin turned the field of natural history upside down with this observation. Until Darwin recognized populations as important, natural historians used a two-thousand-year-old method of viewing and characterizing nature. Aristotle established in his many writings on natural history that things in nature had essences that allowed humans to identify them. Even things that aren't natural have essences about them that allow us to identify what they are, and this identification is extremely important in how we as humans deal with the world around us. For instance, if one searches images of chairs on the internet, a wide array of styles of chairs will appear, ranging from standard folding chairs, to magnificently designed chairs by Frank Lloyd Wright, to bean bag chairs, to chairs designed and built in the Middle Ages, to Chairy, Pee-Wee Herman's talking chair in his playhouse. All of these chairs have an essence about them (for lack of a better word, we can call it "chairness," or, if you are Pee-Wee Herman, "chairiness") that allows us to give them all the general name "chair."

So it is with organisms, as Aristotle clearly articulated in his natural history writings. The essences of things allowed people interested in natural phenomena to easily classify them. Through this typological process things in nature were reduced to commonalities that existed among them. In a mathematical context, this meant that people interested in nature were only looking at averages or characters of the types. Any variation around the average was ignored; this codified the typological approach to understanding nature prior to Darwin. Today, we still use typological approaches in the natural sciences and in evolutionary biology specifically to classify things. We also use the typological approach in systematizing the relationships of organisms to each other.

Biologists who study Darwin, like Harvard zoologist Ernst Mayr, point out that Darwin turned this typological way of thinking upside down by insisting that the type was an unreal, unnatural construct. But since types or averages were unreal, and, more importantly, because they ignored variation, the typological approach ignored something very important about nature. According to Mayr, it took Darwin's focus on variation to break this age-old way of thinking down and allow for a new way to approach nature. Without the focus on variation, natural selection would probably have remained hidden from scientists. Variation is of tantamount importance in understanding the processes that occur in evolution, and without incorporation of variation into our modern way of thinking about nature, we would not be able to interpret the wonderful diversity of colors in nature. Color became a beautiful, wonderful, and technically tractable way to characterize natural selection in action.

Darwin also recognized that organisms often produce many more offspring than necessary for survival, and that many of these offspring are fated to die out before reproducing. Two things are important here with respect to color. First, because there are many, many more organisms produced than can really survive to the next generation, very slight variation in these organisms becomes an obvious way for these populations to be winnowed to their proper sizes. This variation was sometimes imperceptible, as Darwin referred to it, and it led to his insistence that evolution occurred very gradually, slowly, imperceptibly.

If that wasn't enough to get him to the holy grail of natural selection, Darwin also recognized that organisms can't produce millions of offspring and have all of them survive because resources are limited. He gleaned this basic requirement for natural selection from reading an essay by Thomas Malthus. Malthus recognized that when resources are limited, populations reach a tipping point where they can no longer support larger population size. Again, here small changes in physical or behavioral attributes, like color or how organisms respond to color, lend credence to Darwin's preference for gradual change as a result of natural selection.

Darwin and Wallace set the prevailing research paradigm for evolutionary biology with their description of natural selection. Darwin recognized that the environment in which an organism lives is critical for natural selection to work. Both he and Wallace recognized that different environments lead to different kinds of organisms, so they incorporated ecology into the evolutionary picture. Just after the publication of their original work, natural historians of the time who read and understood them scoured nature for examples of the process of natural selection. In addition, mathematicians and statisticians tailored their trades and honed their tools to address evolution and natural selection in the early to mid-1900s. The giants of evolutionary study in the 1940s (Ernst Mayr, Theodosius Dobzhansky, and G. G. Simpson) declared that a Modern Synthesis of evolutionary biology had been established by the vast interest in natural selection both in nature and in theory around the principle. Wallace and Darwin indeed set the stage for over a century of research in color adaptation in nature.

Black and White

There are many examples of natural selection using color, some more notorious than others. H. B. D. Kettlewell's example of industrial melanism is both illuminating and notorious. It was once called one of the most beautiful experiments in evolutionary biology. He took advantage of a tried-and-true ecological evolutionary approach called "mark and recapture." The idea in this approach is to manipulate the environment and the populations in nature to make some inference about the organism's response to the environment. Consider an experiment where one hundred rabbits marked with a red dye are released into two different environments. How many are recaptured will tell the researcher something about the hazards of the two environments with respect to the rabbits. So, all one needs is a method to mark the subjects of a study and two different environments at hand.

Kettlewell noticed that a particular moth in England, named *Biston betularia,* or the peppered moth, varied in its coloration at the most basic level. Some moths were black in coloration and others were white, with a third form sort of in between. He also noticed that the black version was found in high numbers in areas where industrialization was extreme. He chose a site near Birmingham, England, where factories would spew out ash and smoke on a daily basis. You see, moths like to sit on trees (or so the story goes) during the day and are active at night. The ashy industrial environment of Birmingham caused the trees to discolor from their normal white lichen-covered color to a deep black color. The black moths could hide easier on the blackened lichens. In places where industrialization was absent or low, like forested regions near Dorset, England, the lichens were not perturbed by pollution, and this is where the white moths were found in higher numbers.

Kettlewell and others before him could have quit right there and evidence for the role of natural selection in molding variation would have been fairly convincing. However, he decided to perform one of those difficult-to-do ecological studies by controlling the number of moths of each kind (white or black) in the two different environments. He correctly reasoned that any decrease in numbers of one color of moth or the other would be due to predation, which he also reasoned was connected to the capacity of the different colored moths to hide in plain sight on the best background. His colleague Niko Tinbergen even got in on the experiment by filming birds consuming moths, which was pretty good proof that birds were a predatory pressure on the moths. He "released" (note the quotation marks for now) large numbers of black and white moths at Birmingham after marking them with little dots of paint. He left them for a while and came back to count how many black and white moths he was able to "recapture" (note quotation marks for now, also here). In Birmingham, the ratio of black to white recaptured moths was about 3:2. This ratio indicated that natural selection in the guise of predatory birds preferred the black moths. When he did the same experiment at Dorset, he found the ratio of recaptured black to white moths was nearly 3:1. He next reasoned that the selection pressure for white moths in Birmingham was less than for the black moths at Dorset. While we won't go into the mathematics of the system, he could calculate this difference in selection pressure and use the data to show some very basic things about natural selection in action.

While the whole peppered moth story is indeed compelling, Kettlewell's experiments have been criticized from many directions. One criticism is that instead of releasing the moths, he had actually glued them to the trees. Another critique suggested that these moths do not prefer to alight on the trunks of trees in the area. The evolutionary biologist and author Jerry Coyne described his dismay upon learning that Kettlewell's experiments were not as clean as taught in biology classes. He compares the experience to his childhood discovery "that it was my father and not Santa who brought the presents on Christmas Eve." Coyne even went as far as suggesting the example be removed from biology textbooks, which in many instances it has been.

Unfortunately, this seemingly wonderful example of coloration differences, camouflage, and natural selection has recently been dug up by creationists who will try anything to discredit evolutionary principles. One might say that Kettlewell tried "anything" too, but this still ignores the initial observations Kettlewell and others made on distributions of the peppered moths, which prompted Kettlewell to do the experiments in the first place. While he wasn't able in the end to really say he had measured natural selection, he did show that the selection pressure could exist in the guise of bird predation on the moths, and that this would indeed lead to evolution of the peppered moth color morphs. As Coyne suggests, maybe it was not a great example of the nitty-gritty of natural selection, but it was certainly a great example of evolution in action. However, there are better ones out there. While this rather notorious example is flawed in some specific ways, other studies using coloration approach the holy grail of clear demonstration of natural selection.

One of the better examples also uses the basic black and white ecological juxtaposition but involves a vertebrate. The rock pocket mouse, or *Chaetodipus penicillatus*, is found in the American Southwest. It has adapted to the background color on which it lives. In this area, old geological formations have produced sandy white places where the mice live quite happily. In some regions, though, recent volcanic activity has led to lava flows cutting through these white sandy areas, producing new habitats with a black background. It is quite easy to show that mice with white coats fare much better in the areas where there are no lava flows and black-coated mice fare better in areas where there are lava flows. The selective force here is raptors, who like to dine on the mice and use their acute vision to find them. A black-coated mouse is very difficult even for a sharp-eyed raptor to see on a black background, whereas a white-coated mouse is much easier to visualize on the black volcanic rock backdrop. Studies have been conducted to measure the selection pressure in this system, and it is relatively large. Knowing the selection pressure in this context allows researchers to make predictions about how rapid evolutionary change will occur. Again, without going into the math too deeply, if the selection pressure on white mice in lava environments is, say, five times more than the selection pressure on black mice in the white sand environments, then it will take more time for the black mice in white environments to "take over" the white environment. But the real kicker of this story is that researchers have determined which genes of the twenty thousand or so in the rock pocket mouse genome are responsible for the polymorphism and how the genes have changed over evolutionary time.

How do you determine which of the twenty thousand or so genes are involved? There are two major approaches to this problem in evolutionary genetics. The first is genome-wide association studies (GWAS) and has only been made possible in the last few years with genomic technology. This approach looks at the whole genome of the organism and asks, Where in the genome are weird things happening? By "weird things" we mean departures from the normal makeup of the population of organisms being studied. This approach takes a lot of work and is prone to finding a lot of genes that may not really be involved in the trait being chased down—a phenomenon called "false positives." It is a great approach when the trait is under the control of lots of genes.

The second approach involves a lot of luck and is called the "candidate gene approach," or "let's go fishin." In this approach, the possibility of getting false positives is practically nil. However, the probability of finding the gene or genes is diminished, because if you aren't clever from the very beginning, you might not include the right gene in your list of candidates. So what is needed from the start for a candidate gene approach is a lot of knowledge about the trait in question. Fortunately for the researchers working on the question of coat color in mice, the biogenesis of mammalian hair, from its basic morphology to its coloration, has been studied in detail. To simplify how this works, we need to know that hair ultimately takes on its color from pigments expressed by genes. The proteins produced by this gene family are called "melanins." There are two major kinds: pheomelanin (red and yellow forms) and eumelanin (black or dark form). The final color of a hair is determined by the ratio of the pheomelanin to eumelanin.

To understand how this works, we need to know a little about the anatomy of a hair. A hair itself is imbedded in the skin. The part that sticks out of the skin is called the "shaft." The shaft is basically what its name implies, and it's filled with pigments. The shaft is connected to a small bulb well under the skin surface. This bulb contains specific kinds of cells called "melanocytes," which is where the melanins are produced. The melanins are then transported via little blobs of melanin called "melanosomes" to another kind of cell called "keratinocytes," which produces keratin, a structural protein for the hair shaft. As the keratin is produced, melanin is incorporated, and the shaft takes on the color of the melanins. If you were an evolutionary biologist looking for candidate genes and you knew this much about hair and hair color, you would have some pretty good ideas about which genes to look for. First and foremost, you might want to examine the genes that are responsible for melanin production—the eumelanins and pheomelanins. Since the ratio of these two kinds of melanins is the ultimate arbiter of hair color, you might also want to look for genes involved in the regulation of the amount of the melanins produced. You might want to look for genes involved in how well the melanins are transported to the melanocytes and keratinocytes. Hairs themselves develop from the follicle into full-blown hairs through growth pulses controlled by genes, so one might also want to look at how these pulses might be involved in the coloration of the hair.

Michael Nachman is a geneticist who worked in the US Southwest on the pocket mouse colat color system. He and his colleagues point out that around eighty genes have been implicated in the development of hair and its coloration. That's still a lot, so Nachman and his colleagues decided to focus on the eumelanin/pheomelanin ratio as a good first guess at finding candidate genes. What a guess! They knew from years of research and molecular characterization of genes involved in hair development that two proteins were implicated in the control of the overall ratio of the melanins in melanocytes.

They determined that two genes are involved and showed it by going on one of the most well-informed and stacked fishing expeditions on record. The two candidate genes are melanocortin 1 receptor (MC1R) and a gene called "agouti signaling protein," or *agouti* for short. It is instructive to ask how the products of these two genes interact to produce a wild type or unmutated mammalian hair—in other words, how the color pattern is produced in the lab mouse. It turns out that the lab mouse has hair with banded coloration. The tip is black, the middle is yellow, and the base is black.

Mice with this kind of colored hair are called "agouti mice." Agouti cats are a bit different in that they have the striped hair shafts, but they also have solid-colored hair shafts interspersed with the striped ones. The relevance here is that black-coated pocket mice are not banded! How the striping happens is now critical to understanding why *MC1R* and *agouti* are good candidate genes.

MC1R is a G-protein coupled receptor or a GPCR. Remember those? We talked a lot about GPCRs in chapter 2 because opsins are in this same family of proteins. Remember, these kinds of proteins are integrated into the cell membrane and are signaling proteins. *MC1R* itself doesn't make a pigment but instead signals the cell to increase the production of eumelanin by regulating the internal milieu of the melanocyte. When *MC1R* is activated, it produces a ton of eumelanin. If the *agouti* gene is activated, it will make a protein that increases the production of pheomelanin and suppresses the production of eumelanin. *Agouti* is a so-called antagonist of *MC1R*.

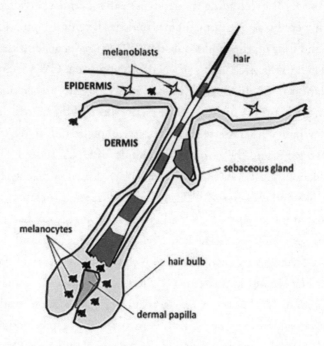

Figure 4.2. Schematic diagram of a hair follicle. *Drawing by Rob DeSalle.*

We need to look at the three-phase production of the hair cell to understand the striped hair pattern. During the first phase, the *MC1R* gene is on full blast. This hair in this phase of development is filled with eumelanin. A pulse of *agouti* expression is regulated during the middle phase, which means that the middle of the shaft is going to be influenced by melanocytes and will contain a lot of pheomelanin. *Agouti* is shut down after the middle phase of development and the *MC1R* protein is now free to regulate eumelanin expression, which results in the base of the shaft being dark again. When Nachman and colleagues looked at the hairs of light-colored rock pocket mice, they observed that the shafts were always banded. When the black-colored rock pocket mice hair shafts were examined, they were discovered to be fully eumelanic. Hence the educated (very, very educated) guess that *MC1R* and *agouti* deserved candidate status was proven correct.

Nachman and colleagues then sequenced these two genes and large chunks of DNA surrounding the genes in nearly seventy individuals, some of which were black and some light, from six different populations, two on lava and four on the lighter substrate. They used a genetic association trick related to the trick used with a GWAS approach. GWAS approaches use variation in populations of organisms to look for association of the variation with a trait. Instead of the analysis extending across the entire genome, as in GWAS, the association tests were applied only to the sequences of the two genes and the surrounding chunks of DNA. Indeed, for one of the populations in Arizona that has mostly black-coated mice (one of the two populations on lava), the *MC1R* gene showed direct association with the black-coated phenotype. In fact, all of the mice at this site had the same four changes in amino acid sequence relative to other sites, and these four amino acid changes are thought to be involved in the efficacy of the *MC1R* protein. The second lava site in New Mexico showed no association of either *agouti* or *MC1R* genes with the coat color. This strange result means simply that some other mechanism is involved in the production of the dark coat color than what is found in the Arizona population. Once again, nature has figured out multiple ways to solve an adaptive problem. While this study was accomplished in 2003, the actual genetic basis of this second New Mexico population still remains elusive today. Natural selection clearly has been a major factor in molding color in this case though, and the actual genome change in the phenomenon is known at least for one of the populations involved.

Not only has natural selection worked within species to produce multiple answers to the adaptation of coat color to environment but indeed other species that live on varying colored substrates have coat color adaptations based in their genetics. Another rodent's distribution and coat color have also been shaped by the geology of where it lives. Glaciers have ripped across North America, creating varying environments with respect to the color of the soil and hence the background on which these rodents (species of *Peromyscus*, or field and deer mouse) live. Not surprisingly, these rodents, which are broadly distributed throughout North America, have adapted to existence on the different substrates by coat color changes. Here three mechanisms of coat color change have been detected by Hopi Hoekstra (one of Nachman's original collaborators on the rock pocket mouse) and her associates. They've pinned the mouse population's adaptation to the light substrate of the Gulf Coast to the interaction of *agouti* and *MC1R*, but the same phenotype on the Atlantic coast cannot be associated with this same interaction; indeed, neither *agouti* nor *MC1R* is involved in the color changes there. In the sand hills of Nebraska, a range of phenotypes has been linked to the lighter coloration of deer mouse populations there. The genetic basis of this light coloration has been attributed to just the *agouti* gene.

By adding this system to the understanding of coat color evolution in mice, we can say there are four distinct possible mechanisms controlling coat color for rodents, all of them independently evolved. *MC1R* alone can be responsible for coat color shift, and likewise *agouti* alone can be responsible. Epistatic interactions of *agouti* and *MC1R* are the third mechanism. The fourth would be an as of yet undiscovered mechanism, which in reality could be several different ones. In other words, evolution has co-opted many different possible ways of accomplishing the same goal. Other examples of well-characterized black-and-white color adaptation, or adaptive melanism, as specialists like to call it, involve lizards and horseshoe hares.

Getting the Rainbow (and More) Involved

Adaptive melanism is perhaps the best studied color phenomenon when it comes to natural selection. And if this was all there was to how natural selection sculpts color variation Darwin and Wallace would not have seen the wondrous things they saw, and this chapter would end here. But organisms in nature do vary widely with respect to coloration, and indeed a literal rainbow of phenotypes and evolutionary phenomena has arisen on our planet. Inherent to that rainbow is the fact that organisms can see a wide range of colors. If all organisms were red-green color blind, our world would be a very different place. But because organisms incorporate a wide range of opsins into their retinas, and can hence detect different colors, we have the wide-ranging palette of organismal coloration that we see every day.

Compared to the rest of life on our planet, we humans respond to a very narrow range of light wavelengths. As we have seen in previous chapters, this range includes wavelengths between 400 and 700 nm. Red is at one end (400 nm) and blue is at the other (700 nm). Over half the radiation of light that comes from our sun is in the IR range (greater than 700 to 1,000,000 nm wavelengths). With all of these IR wavelengths around, how have organisms dealt with it? Infrared radiation comes in two major flavors—near and far. Near infrared radiation (NRI) is radiation that is very close to the limit of visible red light waves. In other words, it is radiation that has wavelengths very close to the 700 nm limit of visible red light our eyes can detect. Far infrared radiation (FRI) can be as long as 1 mm in wavelength, slightly shorter than microwave kinds of light radiation. Some organisms on this planet have evolved to detect NRI.

The neurological information that organisms "see" at NRI wavelengths is hard for us to conceive of, and any attempts on our part to visualize what is observed at these longer wavelengths are fabricated at best; they're actually called "false images." Pit vipers can sense IR signals coming from objects in front of them. They use an organ called a "pit," which exists on the side of their heads between their nostrils and eyes. The pit organ is enervated by nerve cells that have a special kind of receptor molecule in them. Like opsins and other receptor molecules, this receptor, which is closely related to a taste receptor called the "wasabi receptor," is embedded in the neurons in the pit. Instead of interacting with a molecule, these receptors respond to temperature. Since they are snuggled into neural cells, when the temperature receptor is tripped, it sends a message to the snake's brain, where it registers as information. The information from vision and from the pit that the viper collects does not overlap into a single overall picture of what is in front of the viper though. So, the viper isn't "seeing" a single image. Rather, one set of information (visual) complements the other (temperature), but they are processed differently, just as the smell of an object is processed in a different way than its image. Just as seeing images using opsins works best in light environments, the pit sensing works best if the object being detected is on a cold background, also showing that the information from the two sensing systems are processed in different ways.

Recent work revealed that humans can detect NIR too, but not in the same way that pit vipers do. Grazyna Palczewska and colleagues showed in 2014 that the human visual system can detect NIR using the opsin system. They were working on NIR laser rays, and some of the researchers noticed a pale greenish hue around the laser that they were not able to explain. The wavelength of the laser was 1,060 nm—well beyond the 700 nm limit of visible light for humans. Previous researchers had reported the phenomenon and suggested that wavelengths up to 1,330 nm could be detected by human rods and cones. Palczewska and colleagues set out to determine what the hue was and how humans could detect it. They first needed to pin down the anecdotal nature of the observation that NIR could be detected by the human eye. Using a series of clever psychological experiments with human subjects, they clearly showed that indeed the human retina could detect light waves in this NIR range, and that oddly enough, when the wavelengths got longer than 900 nm (200 nm greater than the visual limit of red light), the signal increased. NIR light of longer wavelengths was detected better the longer it got. Next they determined that it takes two photons of NIR wavelength to activate the pigment proteins in the retina. The very same cells (rods and cones) that detect visible light detect NIR light, unlike the viper pit cells detecting NIR using a completely different receptor system. In this case, the visual light and NIR light do not complement each other neurologically but rather overlap each other—hence the information from the two appear to overlap.

So much for light with wavelengths in the infrared. What about other light? As far as we know, organisms cannot detect FIR light (longer wavelengths than NIR but only up to about 1 mm), so no organism detects microwaves (1 mm to 1 m) or radio waves (greater than 1 m). Perhaps organisms simply do not need information on light in this range, but it is rather clear that organisms on our planet do not utilize information from light in this range. We humans have learned to take advantage of radiation at these wavelengths, though, with our microwave ovens and use of radio waves for communication. We don't so much detect these wavelengths as exploit them. Light with wavelengths shorter than 400 nm can however be detected by a wide range of organisms on this planet. How they do it is very logical and reasonable and, most importantly, the result of the evolutionary process.

There are two watershed moments in the history of our understanding of UV light detection by organisms on this planet. The year 1881 is the first watershed moment, when Sir John Lubbock performed a series of charmingly described experiments with ants, bees, and wasps to show that these insects responded to UV light (one section of his paper describes, perhaps wrongheadedly, the "Affection and Kindness" of ants). It was a watershed because the prevailing dogma about animal color vision was that all animals on the planet basically used the same range of colors without variation and this range did not include UV light. The second watershed moment started exactly a century later, in 1981, when UV vision in birds, reptiles, mammals, and other organisms began to be documented. The next decade saw a flood of scientific descriptions of UV vision in a Noah's ark of organisms and culminated in a 1992 review by Gerald H. Jacobs on the extent of UV vision in animals. Since then the literature has been flooded with observations that approach superhero proportions of organisms with UV vision.

We are all familiar with those wonderful photos of flowers in sunlight where remarkable lines and other patterns magically appear under UV light. Too bad humans can't see these patterns directly. Or can they? As we described in chapter 2, the light from the sun (which includes a good amount of UV light) gets focused on the retina by the lens of the eye. The lens is impervious to UV light, so in essence, no UV light gets through to our retina, as it is blocked by the lens. But humans who have their corneas and lens removed during eye surgery do receive UV light onto their retinas and do indeed perceive and process UV light. These humans and indeed other organisms with the capacity to process UV light do so the same way visible light is processed. Their retinal opsins are responsible for capturing photons of light at UV wavelengths, and the cellular transduction process we described in chapter 3, is then triggered to transmit the information to the brain. For humans, the SWS1 opsin for very short wavelength light (near 400 nm) interacts with the photons at these wavelengths, and the unusual information from light of this wavelength is transmitted to the brain and processed along with information from light of other wavelengths. For some organisms that can process UV light, though, novel opsins have evolved to better collect (through more efficient capture) the photons at these lower wavelengths.

We don't see most of the UV light that gets absorbed either; it simply disappears and heats the object a little bit. However, there are some objects that react with UV light and then produce visible light. Such materials are called "phosphors." They can absorb the UV light and reemit it in the visible range of light, some of it at specific wavelengths, where it will be visible as discrete colors. Some phospors emit at many wavelengths, and when added together they can even appear as white light emitted by the object. Hence the wonderful whiteness of your teeth at a club with UV black-lit ambience. What we have just described is artificial fluorescence. There is also natural fluorescence, which we will return to shortly in this chapter.

In summary, humans can extend their light-sensing capacity, but not by much, and when we do it is usually through mechanical assistance or by accident. In many species evolution has created different visual systems with a wider range of wavelength coverage. Are we missing out by having only three opsins and the ability to see only a limited range of colors?

Organisms on this planet have evolved a plethora of ways to exploit a wider range of light perception than our paltry 300-nm range, from 400 nm to 700 nm. Our limited human trichromatic range, by the way, isn't so paltry when compared to most organisms (like most of our primate relatives), who see the world dichromatically. But compared to some of the organisms out there, we are indeed on the low end of wavelength coverage. There are two ways organisms outdo us in their acuity for color vision. First, they can go outside our 300-nm range of visible colors by evolving new opsins that are efficient at capturing light outside this range or, as the pit viper has, by evolving novel receptor mechanisms outside the opsin system. The second way to increase acuity is to increase the number and range of opsins that capture photons and transfer information to the brain for processing. This latter approach seems to be an important mechanism for expanding visual acuity in a wide range of organisms.

When we say evolve receptors outside the range, we do not mean that the organism simply conjures up these new opsins but rather develops them through well-known molecular processes called mutation and gene duplication. Today, typical bacteria have around two thousand to four thousand genes in their genomes, while most higher eukaryotes have on average twenty thousand, which is about the number we have in our own genomes. If we and bacteria have the same common ancestor, how do those numbers work out? It is well known that the genomes of the first organisms on this planet had only a limited number of genes. So the ancestral genome more than likely was also quite limited with respect to number of genes. Higher

In March of 2020, the American Museum of Natural History opened a show called "The Color of Nature." The color photos in this book were kindly provided by the AMNH. Color photo credits go to Dennis Finnin and Roderick Mickens of the AMNH. The exhibition has five galleries—white, red, green, blue, and yellow. Each gallery covers a specific aspect of color and color vision.

COLOR FIGURE 1. Entry. White light versus pure yellow light. In this exhibit the room is bathed alternatively in intense yellow light of wavelength 560–590 nm (left), and strong white light of all wavelengths (right). In the left panel the wall is under intense yellow light; the wall absorbs yellow light in the dark stripes and reflects yellow in the lighter stripes. When the room is bathed in white light you discover that the panels on the wall are actually colored; in this case green and orange. A discussion of the makeup of light is in chapter 1.

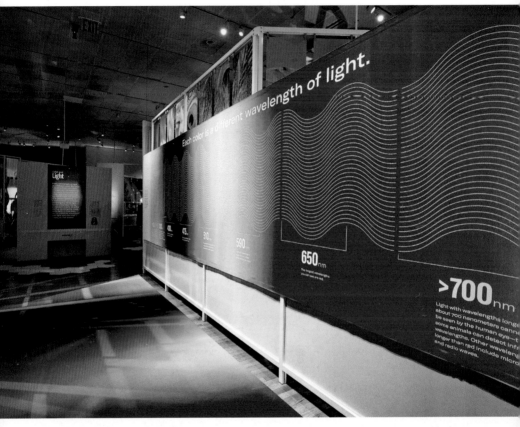

COLOR FIGURE 2. The visible spectrum walk at 1,000,000 times magnification. The walk is 4 meters long, representing the 400 nm over which visible light spans. Each cm in the walk represents 10 nm in the visible spectrum. So, for waves in the low red range, the waves shown have 65 cm wavelength. For a discussion of the visible color spectrum in this book, see chapter 2.

COLOR FIGURE 3. The Green Room Tableau. The exhibit shows enlarged models of several organisms that have evolved interesting coloration patterns. The flower models in the immediate foreground tell a nifty evolutionary story. The large flowers are models of columbines in white (*Aquilegia pubescens*), purple (*Aquilegia vulgaris*), and red (*Aquilegia canadensis*). Different pollinators represented by the large white-lined sphynx moth (*Deilephila lineata*), which pollinates white columbines: the buff -tailed bumblebee (*Bombus terrestris*), which is attracted to blue; and purple and the ruby-throated hummingbird (*Archilochus colubris*), which is attracted to red columbines. The evolutionary reasoning for these associations are that the bumble bees prefer the blue and purple flowers and start the chain of interactions. Since the bees do not go for the red columbines, the hummingbirds focus on red flowers and avoid the already picked over blue ones. The moth is nocturnal and so any flower color that is easier to see at night (ie white ones) are the focus of the moths' attention. For a more detailed discussion of the evolutionary aspects of color, see chapters 4 and 5.

COLOR FIGURE 4. Live animals demonstrating aposematism and camouflage. Left: An example of coloration in mimicry—Henkel's leaf-tailed gecko (*Uroplatus henkeli*) sitting on a tree trunk. Right: An example of warning coloration (aposematism) the Strawberry poison frog (*Oophaga pumilio*). For a discussion of animal and plant coloration in evolution, see chapter 5.

COLOR FIGURE 5. The Red Room. This gallery celebrates the color red by demonstrating its versatility in a cultural context. From left to right the mannequins are Alabama Crimson Tide football uniform, Indian Sari, Catholic Cardinal's cloak, the pink suit, and last but not least the pinkpussy hat. The dress in the center was designed specifically for the exhibition by Brandon Maxwell. The cultural social significance of these colorful clothes is discussed in the text in chapter 6.

COLOR FIGURE 6. The Blue Room. This gallery contains several examples of the use of colors by humans and how humans have created different dyes and pigments over the ages. The interactive in the foreground simulates the production of indigo dye using plant material from the genus *Indigofera*. The interactive goes through the various steps in the indigo dye production. The rest of the interactive is a detailed exhibit on indigo dyed materials and the process of creating colors. For a discussion of the cultural context of color, see chapter 6.

COLOR FIGURE 7. The Yellow Room. This gallery celebrates color's impact on our emotions. The major exhibit in the gallery is a game emceed by Hue, the character shown on the screen. Several questions are asked of the visitor and these are tallied by the interactive to show the visitor the results of the survey. Visitors also interact with the survey in the room on their handheld devices. For a discussion of color and emotions, see chapter 8.

COLOR FIGURE 8 (ABOVE). Playing with color. This interactive allows visitors to develop their own color creations. The visitor can set their own palette of colors and the forms they want to see on the "canvas" in front of them. Their motions then control the flow of colors over the canvas. **COLOR FIGURE 9 (BELOW).** Installment of some of artist Angélica Dass's Pantone portraits. Dass has taken over 4,000 photographs of humans to show the wide variety of human skin color. Each photo is matched to an individual color on a pantone scale which is then listed below each photo. For a discussion of the implications of skin color in race, see chapter 7.

eukaryotes experienced an increase in the number of genes in their genomes. There are many ways to increase the number of genes in a genome (and indeed just reverse the processes and you will lose genes too). An organism can acquire new genes by sloppy transfer. The way this process works is for one organism to shed some of its DNA and for a second organism to pick up the DNA and incorporate it into their genome. This is the main way that bacteria alter the genetic content of their genomes, and it has become a major object of microbial research in the age of genomics.

This is not, however, how most of the changes in the number of opsins have occurred in eukaryotes. The are several ways that eukaryotes can change the number of opsins in their genomes. The first of these involves the duplication of whole genomes during the reproductive process. Such duplications have occurred on several occasions during the divergence of vertebrates and are rampant in the evolution of plants. For instance, during the divergence of animals like insects and vertebrates, two major whole genome duplications are thought to have occurred. Another road to increasing the number of copies of a gene in a genome is when a gene on a chromosome makes two copies of itself through a process called "unequal crossing over." When genomes replicate during reproduction, they usually line up in register, so that only one gene is replicated per the genes in the replicating genome. But sometimes they will line up in unequal register, which will result in more than one copy of the gene being replicated. The flipside of this process, which inevitably happens, is that one fewer gene sometimes gets replicated. It's kind of like an LP, where a scratch causes repeats in some cases or skips in others. When a genome is duplicated, the elements in the new genome are often free to take up novel functions. Likewise, when a gene is duplicated by unequal crossing over, the duplicated gene is also free to take on new functions, either completely different from the original gene or sometimes similar too but an extension of the old gene. We will see in chapter 6 that this is also the mechanism by which the red-green opsins increase and decrease in number in human genomes. Variation in the number and structure of the red-green opsins in humans indeed occurs as a result of unequal crossing over of these genes that are in tight proximity with each other on the X chromosome. Such gene number variation is involved in red-green colorblindness.

These fluctuations in gene number have led to some pretty spectacular end products regarding opsin genome number in animals. To our knowledge the "record" for most opsin genes in an insect genome is sixteen, in what is called a stromatopod, known more commonly as a "mantid shrimp." These remarkable color visualizers have more than five times the number of types of opsin genes in the average human. Coming in at a close second in the opsin gene number sweepstakes is the common bluebottle butterfly (*Graphium sarpedon*). Considering that our three kinds of opsin genes allow us to see millions of different colors, why would this butterfly and the mantid shrimp need fifteen and more? Do they really need to see more than a million colors to survive?

Kentaro Arikawa and his colleagues, who discovered the existence of these fifteen kinds of genes in the bluebottle, suggest that the butterfly actually uses only four of them to interpret color in the environment. Indeed, the researchers who study the mantid shrimp also suggest that this animal does not use all sixteen types to interpret colors. Arikawa and colleagues suggest that these extra opsins enhance the butterfly's capacity to detect other stimuli that organisms and objects around them emit. One obvious stimulus is UV light—something we have previously discussed. But they also suggest that the opsins are used to detect fast-moving objects (like predators) against solid color backgrounds (like a blue sky). The mantid shrimp researchers also point out that this animal has opsin proteins inserted into tissues other than eyes. These extraocular opsins more than likely detect light, but because they are embedded in non-ocular tissues, they have evolved different capacities for detecting light and hence different molecular characteristics.

Another possible explanation for the large number of opsin genes in these organisms comes from knowing that they have much narrower response ranges than our opsins. While our opsins can capture photons over long wavelength ranges, the opsins in these organisms have very narrow capture ranges, meaning two things in general. First, the colors they detect are very specific whereas our opsins detect a large range of colors. Second is related to how we interpret the overall "final" of an object. Remember that our brains "calculate" the amount of capture of different photons by our cones to give the final interpretation of color. The mantid shrimp and the bluebottle butterfly would have to do less calculation, since there is less information coming into the brain from each cone cell. In essence this might also mean that even though the butterfly and mantid shrimp have more opsin types, they might not see color as well as we might think for that number of opsins. Indeed, researchers have determined that their color acuity is not as extreme and broad as it could be if the individual opsins captured a broader range of photon wavelengths. The advantage is that their brains do not have to do all the calculations ours doso they might be able to respond to color stimulus more quickly and without as the level of energy expenditure that our brains do.

For vertebrates the opsin gene number sweepstakes stops at about fourteen, but they add a new slant to the story. The winner for vertebrates is the spiny silverfin (*Diretmus argenteus*), a deep-sea fish with an unusually large eye and a daunting frown on its face. The young of this fish swim in rather shallow water and only make proteins from a few of the fourteen potential opsins. But the adults, who live in much deeper parts of the ocean, over two thousand feet down, produce a whopping fourteen of these genes. The twist is that most of them are rod opsins and not the cone opsins we have focused on for most of this chapter. All of this makes sense if you remember that rod opsins are the retinal workhorses of seeing in the dark. In 2019, Zuzana Musilova and colleagues examined the biology of this deep-sea fish and determined that the multiple rod opsins do indeed cover the wavelength ranges of the various kinds of light that this species encounters two thousand feet below the surface of the ocean.

Beyond evolution

We humans have found ways to extend our vision and the enjoyment and interpretation of it well beyond the limits of evolution and our biological senses. We have developed, over centuries and with ever increasing speed, new materials and gadgets that allow us to create and see many more color effects, to interpret the color images around us in more and subtler ways, and to create color images that in some cases rival nature.

In regard to "seeing" outside the visible, as any child of the '60s knows, a black light (which is a strong source of UV light) can bring out some mind-blowing color effects when directed at objects. In danger of letting the reader know our ages, there was nothing as cool to us as going into the local coffeehouse in college and having our mind bent by the many black light posters on the wall. The modern equivalent is the crazy UV light–induced colors observed at raves, clubs, and other gatherings. What do humans see when viewing things under UV light then? The simple answer is visible light! A black light emits tons of light from the UV range. That light appears dark purple to our eyes because the black light also emits a tiny bit of light in the 400-nm range, which our opsins can handle. But the majority of the light is absolutely invisible to our eyes. The UV light starts to hit objects and is absorbed and reflected by the objects like any other light. Again, we do not see the light that is reflected because the UV waves reflected are simply not detected by our opsins.

At the other end of the spectrum, we humans can detect and interpret visible light and colors with the opsins embedded in the membranes of our retinal cone and rod cells. Infrared waves are detected by a completely different kind of receptor molecule, which isn't embedded in the eyes. Any attempts on our part to visualize what is emitted at the long infrared wavelengths is barely possible.

However, we now have cameras that can detect IR with a reasonable spatial resolution, either as special devices or even as add-ons for our smartphones. When such IR cameras are used by humans to view things in the IR wavelength regime, we can only recognize the color interpreted on the screen, not the wavelength of the light. At best we can take the intensity at one or several different IR wavelengths and artificially turn this into "false color images." For instance, the intensity of infrared light is commonly visualized using heat-based colors: red for hot, blue for cold, and other colors in between. Why is this? Objects that are cooler emit more light in the Far Infra Red (FIR), less in the Near Infra Red (NIR) and visible spectrum; hotter objects emit a large amount of light in the visible and NIR ranges. Our human way of deciphering these readings of temperature is to use a thermal range of false colors. In contrast, an organism that senses IR doesn't "see" these thermal false colors, as it doesn't use its eyes to detect them. Such organisms will get the impression in their brains of a far extended rainbow, with colors for which we have no words.

There are many technical refinements that have expanded our color vision beyond the paltry 300-nm range that natural selection handed us. We now have cameras, for still photography and for video, that can view very specific wavelengths. These can be associated uniquely to the chemical elements, or molecules, in the objects we are looking at. Having several images as separate wavelengths—NIR, visible, and UV—combined gives us an enormous amount of information about the object. In recent years we have been able to generate such images of the sun and make videos at many different wavelengths. This technology allows us to create a very detailed model of the composition and dynamics of radiation emission from the sun. We can detect the composition, the amount of hydrogen, and other elements in the sun surface. We can view the movement on the sun's surface, measure the temperature, see patterns such as flares, and reconstruct the magnetic fields on the sun's surface and much more. The vast majority of the information we have about the sun and others solar bodies comes from such spectrally resolved images. Using similar technology, we can look at the human body, and inside the body, with great detail. Simple applications used in airports find people with elevated temperatures, who might be infected. More complex techniques can be used very efficiently to diagnose various diseases. We have turned our extended color vision into a great tool for analyzing us and our environment. We have, in a sense, expanded our color vision beyond the normal range that natural selection set for our species during our evolution.

5

Gary Larson's
Animal Coloring Book

Recently, a group of color vision biologists published a discipline-defining review articles in the journal *Science*. In this 2017 paper they outline the future of color research in natural history studies. Innes C. Cuthill and Tim Caro, along with over twenty other colleagues, suggest that aposematism (warning coloration), crypsis (camouflage), sexual selection, self-producing light (bioluminescence, fluorescence, and light shifting), social constraints, and contingency are all involved in color evolution in nature, and this should be where the action is focused in color research. If some of this sounds familiar, it more than likely is, as it should harken back to the last chapter, where we discussed Alfred Russel Wallace's prescient outlining of color research at the dawn of Darwinism. But there is another figure in biology who has also touched on most of these topics. The cartoonist Gary Larson has been able, over his illustrious career, to add context to science with simple cartoons. He has focused many of his creations on animal and plant traits that add to the organisms' success in nature.

Aposematism

One of our favorite and most memorable cartoons relevant to aposematism is entitled "How nature says, 'Do not touch.'" Since Larson did not draw this cartoon in color, he didn't address warning coloration, but he does capture the essence of warning mechanisms that animals use to stave off unwanted contact or interest. He shows a rattlesnake rattling, a pufferfish puffing, and a cat being a cat in three of the panels. He then humanizes it up a bit with the last of the four panels of the cartoon, where he shows a rather disheveled human male in a trench coat, stuck in a rubber animal inner tube, holding a bazooka and standing on an alley corner. Yes, stay away from that last one for sure.

Color has evolved in animal and plant systems to warn off, avoid irritating, and prevent possibly fatal contact from other organisms. Larson's cartoon explicates the major assumption behind such warning. Some conspicuous visual, olfactory, or auditory cue is processed or interpreted by an organism as threatening, and it learns to link danger and contact with such a conspicuous individual. Whereas Larson's human danger sign with the bazooka is a single individual (we would hope), most aposematic systems rely on the entire population, or at least one of the sexes, carrying the warning. This reliance reinforces the warning signal to predators. Also inherent in any aposematic system is the fact that the organisms bearing the warning coloration possess something dangerous that they can do, unless they are of course mimicking (see below) a phenotype or pattern that is indeed toxic or dangerous. The potential danger of a bazooka-bearing, inner-tube-wearing alley dweller is pretty evident to us humans, but two black spots on an orange background on the carapace of a small beetle aren't that scary to us at all. But this pattern can pretty efficiently ward off bird predators, because they know this ladybird beetle carries a potent poison.

Biologists use some bizarre methods to measure things in nature. Perhaps some of the most bizarre assays test poisons or distasteful compounds that predators consume during their foraging in order to discern their effects on predators. One of the most famous comes from classic work on the monarch butterfly. Lincoln and Jane Van Zandt Brower pioneered this work by realizing that the orange coloration of the monarch was in some way warning off predation. They did not have the tools to figure out the ins and outs of the natural system, so Lincoln Brower created a whole new field of biology called "chemical ecology." He needed an assay that would allow him to characterize the toxic nature of a butterfly's body, so he developed what he called the "emetic unit" (in common parlance this would be called the "puke unit"). Brower force-fed a powdered concoction of toxic monarch adults to blue jays. He then observed the jays' reactions to the forced meal and could characterize the violence by which the jays puked out the toxic powder. With this assay he was able to characterize different populations of monarchs for their relative toxicity. He also used this assay to establish a connection between the milkweed plant and its toxic cardenolides, which monarch larvae feed on, and the ingestion of the cardenolides, which could then be sequestered into the adult bodies of the butterflies. These results became the basic facts needed to establish the true nature of the warning coloration of monarchs.

Ladybird beetles have long been a focus of natural history study when it comes to coloration. Darwin was an avid beetle collector, and ladybirds were among the targets for his beetle collections. There are over four hundred species of ladybirds, and many of them are brightly colored, tipping their hands as being aposematic. The larvae of some of the more conspicuously colored ones eat pesty aphids, mites, and other other insect eggs, so these beetles have become the darlings of the gardening culture. The stuff they eat can be turned into very smelly chemicals that are toxic to other insects and small vertebrates. These beetles are usually not toxic to humans unless consumed in large numbers—one estimate is that about 100 ladybird beetles would make a grown human pretty sick. Because of their distinct coloration and stinkiness, they have been studied recently as an excellent example of aposematism.

In an equally creative, but perhaps more deadly, assay with respect to emetic units, Lina María Arenas, Dominic Walter, and Martin Stevens measured the toxicity of several kinds of ladybird beetles. By placing the bodies (sans elytra) of individuals of six different kind of ladybirds into tubes with methanol and crushing them to release any molecules in the carcass, they created cocktails of ladybird toxins for the various species they were testing. They then placed carefully measured concentrations of the cocktails into reaction vessels containing *Daphnia pulex,* or water fleas. Poor *D. pulex* have gotten the reputation of being good monitors of ecotoxicity over the years because they die in the presence of toxins in a very constant, measurable, and comparable way. You simply count the number of dead water flea bodies floating in the aftermath of the experiment to get a measure of the relative toxicity of ladybird bodies. In this way, the scientists were able to measure the relative toxicity of six kinds of ladybirds—the larch, orange, pine, 14-spot, and two kinds of 2-spot (one with the usual orange background and two black spots, and a so-called melanic form with the reversed black background and two orange spots), and rank them with respect to the power of their toxins. Sounds like a pretty cool, cruel experiment, but there was also some method to their meanness. They wanted to compare this level of toxicity with how true or "honest" the aposematic coloration signal was.

Natural historians had for a long time recognized that some aposematic organisms carry a lot of color variation in their populations. And indeed, there is a lot of variation between species and the intensity of their aposematic signals. But this observation is counterintuitive, because the reasoning goes that it is better for a toxic organism to have very little variation in warning coloration in order to reinforce the pattern on its predators more efficiently. As all paradoxes go, this one begs exploration. Arenas and her colleagues found that they could arrange the ladybird species in the following manner from least toxic to most: larch < 14-spot < pine < 2-spot normal < 2-spot melanic < orange. They could then measure the intensity of the aposematic signal and see if the intensity (honesty) of the signal correlated with the toxicity. They concluded that "signal contrast against the background is a good predictor of toxicity, showing that the colors are honest signals." They even made tiny model ladybirds painted with varying signal contrasts to test natural populations of bird predators. Their model experiment showed that natural predators also avoid the more highly contrasted coloration patterns, completing a pretty amazing story about the honesty of warning coloration.

Aposematism is found broadly in insects. Here we have looked at two of the orders—Coleoptera and Lepidoptera—but nearly every one, if not all, of the twenty-nine orders of insects has some form of aposematism in some of its species. Aposematism also exists in vertebrates, plants, and many non-insect vertebrates. No wonder Wallace was so tuned into it, and no wonder it tipped him and Darwin off to the prevalence of natural selection in nature.

Mimicry

It stands to reason that since natural selection doesn't have a conscience, any and all strategies for survival will eventually get a chance. Cheating has become one of the best and most "economical" strategies of them all. Cheating is simply a way of life on our planet. Why wouldn't one species take advantage of the difficult chemical properties of another species? Looking like something dangerous, toxic, poisonous, or smelly without being so is a major cheat strategy that has been adopted by millions of species in the history of our planet. Sometimes it is more economical in an evolutionary sense to look like something dangerous than to make yourself dangerous, so mimicry gets a foothold in many ecological systems. Mimicry can also include looking like something else that predators will avoid because of disinterest, like a dead leaf or a stick. And coloration has been a major factor in all kinds of mimicry.

We have already mentioned mimicry in the context of the monarch butterfly. Again we return to Larson, who refers directly to animal mimicry in many of his cartoons; his best cartoon on the topic is titled "When animal mimicry breaks down." It shows a hunter dressed in full winter hunting clothing, kneeling on the ground with his trusted rifle. A tall figure that is obviously a forty-five-point buck dressed in hunter attire stands behind him, annoyingly saying, "Howdy! Any luck? The vacuum bag is hot today! Any luck? Howdy!" Apparently, the reference to the vacuum bag gives the poor buck's mimicry of a hunter away. Mimicry is only effective when there is a model and the mimic is as true to the model as possible.

Here we will discuss two model mimic systems that cover a lot of mimicry ground: milk snake-coral snake mimicry and *Heliconius* butterfly mimicry. The first because some of the genetic machinery behind the color patterns in mimicry is known, and the latter because it is somewhat imperfect and through its imperfection teaches us a lot about how color mimicry works in nature. Before going to these amazing animal systems, some terminology is necessary. In the decades following the introduction of the idea of natural selection, many novel ways of viewing nature arose. One naturalist named Henry Walter Bates made trips to the Amazon right after *Origin of Species* was published. He described in detail several cases of similarly colored insects in Brazilian rain forests. The phenomenon of an obnoxious or poisonous species being copied by a harmless species was affectionately named after Bates and is called "Batesian mimicry," in honor of his pioneering work. Theodor (Fritz) Muller, a German natural historian, came along a decade or so later and recognized that in many cases, an aposematic species would be copied by another species that also developed a toxic taste. This form of mimicry was given his name and is now known as "Mullerian mimicry." Today, naturalists refer to the grades between Batesian and Mullerian mimicry as Quasi-Batesian and automimicry. Quasi-Batesian refers to the recognition of a kind of mimicry system that at first appears as Mullerian (i.e., the species in the interaction are harmful) because all species in the system have some level of distastefulness. Instead of there being a broad and equal conference of advantage, some species involved in the interaction are at a disadvantage because the predators learn more slowly about the distasteful or toxic nature of the species involved. Some species in the system are protected nicely and efficiently, while others are more prone to predation, creating a system that is not entirely Batesian. Automimicry refers to the phenomenon within a species where there is variation in the level of toxicity or distastefulness. In this kind of mimicry, some members of the species "cheat" other members of the species. As with Quasi-Batesian mimicry, though, if there are high levels of automimicry, then the predator avoids learning rapidly or doesn't learn altogether. How these different forms of mimicry grade into one another should be obvious, though the level of cheating in the game of mimicry is obviously worse than sandpapering the ball in Australian cricket or stealing signs in American baseball.

A causal trail needs to be discovered and explained to nail down phenomena like Batesian and Mullerian mimicry. When the topic of mimicry arises, the monarch butterfly and a not-so-closely-related butterfly, the viceroy, are always brought up. The viceroy, it is said, is not distasteful, so by looking very similar to the monarch, it was heralded as a wonderful example of Batesian mimicry. But remember Lincoln Brower, the inventor of the emetic unit? One of the major follow-ups to his work on monarchs was to see how pukey viceroys—a mimic of the monarch—make birds. He and David Ritland clearly showed that the viceroy is indeed toxic, having an even stronger emetic effect on bird predators than the monarch does. This discovery changed the lectures of hundreds of evolutionary biology teachers over the globe but pointed to the fact that the different kinds of mimicry described by scientists did not have strong boundaries but rather graded one into another. However, there are good, strong Batesian mimicry systems involving color, and this brings us to butterflies and snakes.

Consider the "Red touch black, safe for Jack. Red touch yellow kills a fellow" rhyme, which is something every human who hunts snakes in the American Southeast memorizes and lives (or dies) by. This rhyme refers to the coral snake (red touch yellow) and the milk snake (red touch black), a pair of geographically overlapping snake groups. Other "red touch black" snakes can be found in the genus *Cemophora* (or scarlet snakes) and *Chionactus* (or ground snakes), both of which are harmless. The milk snake (*Lampropeltis triangulum*) is a non-venomous type of snake commonly known as a "king snake" (actually, it has more aliases than a 1920s grifter), which grows to be somewhat big (two to four feet in length). It is a favorite snake for collectors because of its capacity to breed easily in captivity, which makes it a desirous species in the pet trade (hence the need to hunt it). The eastern coral snake (*Micrurus fulvius*), on the other hand, is quite poisonous, producing a potent neurotoxin that acts directly on muscles essential for lung contraction. It can cause slow but effective lung malfunction and even death if not treated with antivenom.

Strong associations of these poisonous and non-poisonous snakes have been hypothesized by natural historians to explain their distribution. While Wallace established that mimicry in snakes occurs in nature in a famous paper entitled "Mimicry, and Other Protective Resemblances Among Animals" in 1867, the coral snake/milk snake phenomenon wasn't articulated clearly until ninety years later. The validity of the Batesian nature of the mimicry was debated for nearly a century. It wasn't until 1981, when Harry Greene and Roy McDiarmid resoundingly showed the relationship was tenable, that the real nature of the mimicry was accepted widely. The mimicry system evolved as a result of predation pressure on the snakes, more than likely from raccoons, foxes, skunks, and coyotes. This pressure was no doubt strong, and the system has resulted in some of the most imaginative evolutionary research on record.

One of the predictions of the Batesian nature of the phenomenon is that the mimic (non-poisonous species like *L. traingulum*) should be strongly associated with the model (coral snake). Alison R. Davis Rabosky and colleagues have constructed evolutionary trees (see chapter 2) of coral snake models and non-poisonous mimics to examine how these two kinds of snakes have evolved. The surprising result of their study is that shifts in mimetic coloration of mimics is strongly correlated with the presence of coral snakes in time and location. It is as if the mimics are following the models with changes in coloration. This demonstrates the massive strength of natural selection to mold coloration in these populations. One of the outcomes of this kind of strong selection in a Batesian mimicry relationship is that the models strongly impact the mimics. But do the mimics provide any push-back selection pressure on the models?

It makes sense that perhaps the models would, in an evolutionary context, attempt to escape the "parasitism" of the mimics. After all, there is nothing more pleasing than to sting a cheat, and having mimics opens the model up to loss or slowing down of learning by predators. What we describe here is a sort of an arms race that was eloquently outlined by Peter Raven and Paul Ehrlich in the 1960s, early in the careers of these two giants of modern natural history. Arms races are exactly what the name implies. One entity develops an effective strategy for defense or livelihood, and another responds with a strategy to avoid (if it is prey) or exploit (if it is a predator) the first strategy. Of course, all of this happens on an evolutionary time scale. Ehrlich and Raven described the phenomenon in the context of insect (Ehrlich) interactions with plants (Raven).

This kind of selection in the coral snake/milk snake interaction has been dubbed "chase-away" selection and is a prediction of a Batesian mimicry system. Because there is considerable variation in the color morphs of both the models and mimics in this system, Christopher K. Akcali, David W. Kikuchi, and David W. Pfennig looked at coral and milk snake populations for evidence of chase away. Bottom line is that they don't find it, meaning that chase away may not be such a big deal in the evolution of coral/milk snake mimicry. Regardless, the mechanisms for this interesting mimicry system provide strong evidence for the existence of natural selection and Batesian mimicry both. One itching, missing aspect of this system involves the genetic architecture of the coloration patterns. While some studies suggest that the red-black pigmentation patterns might be controlled genetically by a simple two-gene Mendelian mechanism, the actual genetic or genomic basis for the coloration is still unknown. Contrast this with the mouse coat color phenotypes and what we describe below for *Heliconius* butterflies.

Before venturing into *Heliconius* butterfly wing patterns, a warning is necessary. While the overall conclusion that science has a good hold on the genetics and genomics of this important mimicry system is accurate, the literature on the topic is voluminous and complex. Here we distill this century of work on the system and apologize in advance if we have glossed over or underrepresented any of it. With that said, we can set the stage by pointing out that within the genus *Heliconius* there are forty-three species, and they are all found in tropical to temperate areas in South America. *Helconius* butterflies are in a family with eight other genera, and the family is included in a larger group called "nymphalid butterflies." The genus *Heliconius* can be divided into two major groups based on molecular analysis (see chapter 2) and on their lifestyles. Three species—*H. erato, H. cydno,* and *H. melpomeme*—have been chosen for much of the genetic analysis. Two of these, *H. cydno* and *H. melpomeme,* are very closely related to each other and considered to be sister taxa or each other's closest relative and also reside in the same major group of *Heliconius*. These two species are equally related to the third species, *H. erato,* but *H. erato* is in the second major group of *Heliconius*.

All species in the genus feed on relatively obnoxious-tasting plants in the genus *Passiflora,* from which they obtain cyanogenic molecules. Like monarchs, their larvae feed on the plants and sequester the cyanogens into their adult bodies. They also can synthesize cyanogenic molecules on their own from precursors of common amino acids. Balancing the two kinds of cyanogenic sources is a complex trade-off within the species of this genus. While their larvae feed on *Passiflora,* species in this genus have complex interactions with other plants as adults. Since all species tend to feed on *Passiflora,* all species carry cyanogens, and this means that any color correlations or mimicry are on the Mullerian side of the scale of mimicry interactions.

Evolutionary works of art, painted on the wings and represented in the genetic makeup that gives rise to the color palette of the forty-three species in the genus, constitute one of the most intricate dances of animal coloration on the planet. The sleek wings of this group, often called "longwings," are pretty much all the same shape. Like most insects, species in this order have four wings. A first pair of wings attached to the animal's second tarsal segment (insects like Lepidoptera have three tarsal segments) is pointed at the end that attaches to the animal and rounded at the end of the wing. This first pair of wings is roughly twice as long as they are wide. They are almost twice the size of the second wing, which is attached to the third tarsal segment. This second pair of wings is about as wide as it is long.

The color palette of these species includes almost the entire spectrum, with a strong reliance on a black background. Oftentimes white and brown are also thrown onto the palette. As with most aposematic species, there is a strong reliance on orange, yellow, and red in the coloring of these warning wings. But the real beauty of the wings lies in the patterns that are found there. The patterns have some general motifs, like stripes in specific locations, and while at first glance it looks like there is little resemblance of some of the species to others, there are indeed general patterns that each of the wings follows. These similarities are what Fred Nijhout called the "nymphalid groundplan" back in the 1990s. A groundplan is simply what one can infer is an ancestral state. Remember that there are forty-three species in this group, and they all come from a common ancestor that had the groundplan for the group. What this common ancestor looked like can be reconstructed from knowledge of how the wing develops and what the possible outcomes of the wing are.

If one lays out all the wings in one place and looks at the overall patterns of color, one is struck with the prevalence of black in the wings and might be led to hypothesize that the groundplan was a black background on which the reds, yellows, browns, whites, and oranges were "painted" over; in other words, there was a black starting canvas. But breaking with convention and with what the eyes were seeing, Nijhout pointed out that white and yellow should be the groundplan canvas, and that black, brown, and red were "painted" onto these backgrounds. To add to this advance, the biologist Lawrence E. Gilbert spent decades crossing *Heliconius* butterflies in his labs in Austin, Texas, to uncover the basis for *Heliconius* wing color. In the later part of the 20th century, he hypothesized that there are three essential elements of *Heliconius* wing color: a window (usually white or yellow); shutters (red or melanic [black, brown, etc.]) that overlay the window; and walls that are on the border of the wings around the window. Gilbert also clarified that there are two elements composing the markings on the wings; until he figured it out, these elements had been lumped together. The elements are where the markings are placed on the wing (position) and how they are shaped (patterns). Separating these two aspects of wing painting was critical in understanding the overall way that the wings develop. His ideas about how *Heliconius* wings are patterned opened the way for clearer genetic explanations for the patterns.

Give similar canvases and the same color palettes to two artists (it doesn't matter what age), and the art produced will more than likely be very divergent. Give the same palette to forty-three artists and the range of art will be astonishing, probably going all the way from portraiture, to impressionism, to abstract nihilism, to whatever art movement one can think of. But with *Heliconius* butterflies, while the patterns are spectacular and the variation breathtaking, there is a great deal of structure to the outcomes. Between related species, convergence of wing coloration occurs frequently. Even between distantly related groups of *Heliconius*, convergence also occurs. Instead of going hog wild with rampant coloration patterns, the Mullerian mimicry involved in the evolution of the group has corralled the outcomes of color patterns. There are several reasons for this apparent lack of artistic creativity, the most prominent being the strong role of Mullerian mimicry and the fact that similar genetic mechanisms underly the color pattern variation in the genus. Such strong Mullerian mimicry will result in color morphs of distantly related species converging on each other and hence reduce the range the color patterning can explore. Another is called "genomic introgression," and this involves mating between members of different species. It appears that species in the *melpomeme, cydno* group of closely related species can interbreed with each other, and such interbreeding maintains the genes for color patterns of one species into the genome of the other. This sort of homogenizes the color patterning of the two species, or at least makes color patterns from one species available for the other. It turns out that color patterns aren't the only phenotypic elements being moved around in this way, as behavioral and other morphological traits are dragged along in the introgression events.

Several genetic loci have been pinpointed in the production of the various color patterns in the group. Early work on changes between *H. erato*, *H. cydno*, and *H. melpomeme* indicated that tens of mimicry genes could be found in the genomes of *melpomeme*, *erato*, and *cydno*. More precise mapping indicated that there are twenty-two genes involved in *melpomeme*, seventeen genes in *erato*, and an overlapping number in *cydno*. Since there is so much going on here, let's take a look at one simple color pattern in the group and how genes control it. The pattern we examine here is the simple red striping or blotching on the wing. This pattern is found in nature in a lineage of *H. erato* called *H. erato hydara* and a lineage of *H. melpomeme* called *H. melpomeme cythera*. Remember that *melpomeme* and *erato* are in two different groups of *Heliconius* and are not terribly closely related. Early genetic linkage studies suggested that a major factor for red patterning is a gene that codes for the protein kinesin, a cool so-called motor protein because it can literally motor along microtubule filaments in the cell. It is also involved in some very basic cellular functions.

Okay, so this cool protein might be involved in red color patterning, but how? It turns out that it is very difficult to make up a good story for how this protein is involved, and there is a good reason. It isn't the only one in the genomic region involved in color patterning in a big way. When more precise genetic studies were conducted, one of the kinesin's neighbor genes, optix, could also be shown to be involved. But not so surprisingly, when the sequences of optix proteins are compared across several *Heliconius* species, it is obvious they are making very similar proteins! It simply can't be that the optix protein structure is involved but rather, as further studies suggest, that the control of the expression of the optix gene is the major factor involved. As with other things in this chapter, this might sound familiar. Remember that the control of the melanic mouse coat color can be localized to a couple of candidate genes and how they are expressed, not what protein is produced. More than likely it is an interaction between optix and kinesin that is doing the trick, but by simple manipulation of this regulatory region the red stripe on black background can easily be added to, or the color of the stripes changed.

Pinning all of these genetic interactions down has been helped along by the ability of researchers to modify the genes involved in color patterning. In particular, the CRISPR-Cas9 genome editing approach has been used in *Heliconius* to pin down genetic mechanisms of color patterning. This genome editing tool has two components: the Cas9 enzyme that can be programmed to cut out a specific piece of DNA, and a "guide RNA," which transfers the enzyme to its destination on the DNA molecule, along with a replacement sequence. All that is needed to get this approach to work is a sequence of the gene being targeted for editing and a sequence to edit it to. Before this approach, experiments where genes could be modified in species like *Heliconius* were either not possible or incredibly tedious and expensive. The approach gives the opportunity to take any gene one thinks is involved in a phenotype—in this case, color patterning—and rapidly and efficiently edit it to see what it does. This approach has been used to test several color patterning loci for their involvement in color patterning, and most tests have verified the involvement of genes like optix. It will not be too surprising in the near future if researchers are using gene editing approaches like CRISPR or more advanced approaches, like the newly developed prime editing approach, to examine color patterning in all kinds of species in nature. The *Heliconius* system is only the tip of the iceberg of examples of color patterning evolution involved in mimicry, but it is one of the most informative and paradigmatic systems in all of biology.

Sexy

The role of variation between the sexes was one of Wallace and Darwin's favorite topics, and apparently Gary Larson's too. In yet another of his hilarious cartoons, Larson shows a young female insect leaving the house while her parents sit on the couch. The mother says, "Hold it right there young lady! Before you go out you take off some of that makeup and wash off that gallon of pheromones." While pheromones are one sure way to attract the opposite sex in the natural world, colors and their patterning are another. Larson pins down the importance of sexual attraction in nature in the cartoon, and in fact, attraction is at the heart of a strong evolutionary force that both Wallace and Darwin recognized as prevalent in nature. It's called "sexual selection."

This phenomenon can be recognized when a species has males and females that look or act very different from each other. Larson's pheromone example is a good one in this context, as it is the female teenage insect putting the pheromones on and not the male. In the 1800s, Jean-Henri Fabre, a French entomologist known for his advanced ability to communicate science to the masses, did an important experiment in the discovery of pheromones with the great peacock moth. He found a female, placed her in a cage, and left her overnight, accessible to the outdoors. In the morning he awoke and observed forty to fifty male butterflies hanging on the cage, begging for access to the female. He described the incident in the following way: "Coming in from every direction and apprised I know not how, here are forty lovers eager to pay their respects to the marriageable bride born that morning amid the mysteries of my study. For the moment let us disturb the swarm of wooers no further." In Larson's insect home described above, Mom was right. But the point here is that the female has the sex-specific capacity to produce the pheromones and the males the extraordinary capacity to detect them.

In many systems that undergo sexual selection, a strange thing can happen. The trait under selection can evolve so as to seem out of control. This is a process that R. A. Fisher, the famous early 20th-century statistician, described as "runaway sexual selection." This process is similar to but not to be confused with, chase-away selection that we described above for snakes. Fisher based his ideas about runaway selection on the following scenario. A sexual selection system is based on female preference and the focus of male phenotypes on it. The better a male fits the female's preference, the more mates he will get. Important items to add here are that sperm are cheap and eggs are expensive, and males will oftentimes not participate in rearing young. Both of these factors will also contribute to a focus on male phenotypes.

Hence there is strong selection on such traits. Such a system can lead to one of the sexes having highly exaggerated phenotypes, such as appendages or exaggerated coloration. One of the more spectacular insect examples is, believe it or not, found in flies. Females of the fly species *Drosophila heteroneura* are typical flies—two wings, six legs, rounded head with large-faceted eyes, a thorax, and abdomen. Male *heteroneura* have the six legs, two wings, abdomens, and thoraxes, but their heads are spectacular. Their eyes are placed out on stalks, and it appears that the longer the stalks, the more preferable a male is. During mating, females congregate into what are called "leks." Birds are major participators in lekking, but the behavior is found in all kinds of animals. Lekking allows the females to observe and evaluate males for potential mating. As mentioned above, they will focus on a particular trait or behavior in the males and use these to decide on mates. It is also important for the male to get into the lek and to remain there for the female to do her bidding, so males will participate in highly ritualized interactions to oust each other from the lek. For *heteroneura,* the interaction is called "jousting," as they will place their heads together (much like rams in butting contests) and interact in that way. A smaller male will almost always leave the lek when confronted by a male with longer eyestalks. The male isn't in like Flynn yet, though, as the female will decide on the overall length of the eystalks herself. In fact, the *heteroneura* stalks can actually be as long as half the length of the male's body. The runaway nature of natural selection, according to Fisher, was that the males will evolve more and more exaggerated traits to obtain mates, resulting in a runaway-train-like process that oftentimes places a strain on the male anatomy, behavior, or physiology. Being sexy can be detrimental to your health.

Color is involved in many sexual selection systems, most notably in birds. Most readers will be familiar with the peacock male and its elaborate tail, which is involved in courtship with the female peahen. And every reader can most likely think of their favorite sexually dimorphic bird example based on color. The range of color of bird plumages is impressive, but some researchers wonder whether birds have used the entire range of colors available to them during their evolution. Mary Caswell Stoddard and Richard O. Prum have characterized this range and indeed show that given birds have trichromatic color vision, they only use about 26 percent of the possible gamut of plumage coloration. They compared this to the plant gamut and conclude that birds are at least as colorful as plants. This limited range of colors is somewhat puzzling, given the breadth of possibilities open to natural selection, but as we discussed above for *Heliconius* butterflies, the evolutionary process can also limit ranges and prohibit the exploration of color combinations.

The way we have described the process of sexual selection may not give females their fair due. Female birds are oftentimes ornamented and elaborately colored as well, especially in the large group of birds called "passerines" (5,831 species). James Dale and colleagues have quantified the colors of plumages of all species in this group of birds. This amazing data set was used by Dale to examine differences in coloration between the sexes of these colorful birds. They found that female coloration was indeed correlated to male coloration, suggesting a tight constraint of color change in one sex with the other. For both sexes, larger passerine species and species living in the tropics tend to be more ornamented or colorful than their smaller and nontropical counterparts. But the most amazing result of this study was that when sexual dimorphism becomes extreme, it isn't because males are getting more and more extreme but rather that females in these species get duller or reduce their coloration more drastically as males get more colorful. While we have focused on sexual selection as a driver of coloration in animals, the large number of studies in Dale's color database argues that there is more to how colors involved in sexual dimorphism arise. As James Dale states during the study of bird coloration, "There is a lot of variation and it got swept under the rug" as a result of our reliance on sexual selection as an explanation for color differences.

Hiding in plain sight

Not all coloration in animals is bright, attractive, or outrageously pretty. Yet another strategy for survival in nature that involves coloration is camouflage, or hiding in plain sight by looking like your surroundings. For camouflage, we point to classic Larson. Harold, dressed in loud stripes and checks, answers the door to find a monster waiting right outside. Lola sits in a chair, nearly invisible in a dress decorated with the same pattern as the wallpaper, lampshade, ceiling, and the chair she's sitting in. The caption reads, "When the monster came, Lola, like the peppered moth and the arctic hare, remained motionless and undetected. Harold, of course, was immediately devoured." The caption alludes to an example of coloration in nature that we have already discussed at length—the peppered moth, which exemplifies only one of the many ways in which species accomplish camouflage or crypsis. Thomas White, Martin Stevens, and Sami Merilaita suggest that there are six basic ways that cryptic coloration or patterning works: simple matching of background, self-shadowing, obliterative shading, disruption, flicker fusion, and distraction. All of these strategies strive to undercut the vision of predators in one way or another using light. Some, like disruptive coloration, distractive markings, obliterative shading, and self-shadowing, trick predators into misjudging location, edges, or the true form of a prey item. Flicker fusion is an interesting phenomenon in that it uses motion (conspicuous in and of itself) to blur colors into backgrounds. Background matching is perhaps the most common of all of these mechanisms, but a close examination of these indicates that they are not exclusive and sometimes occur concurrently in animals attempting camouflage. As with aposematism and mimicry, camouflage has been touted as one of the prime examples of natural selection in nature.

Fluorescence, bioluminescence, and other light phenomena

So far, we have discussed natural color examples where light is directly reflected off an organism. However, any organism can also bend light, diffract it, redirect it, and, of all things, create it. One of the best examples in this final category of color in nature that we discuss here is bioluminescence. This natural phenomenon is the subject of a Larson cartoon of a pair of insects sitting in their living room looking out their picture window. Two large glowing lights that look a lot like the headlights of an oncoming car appear in the front window of the house across the street. The female insect in the first house blurts out to her mate, "Nik, the fireflies across the street—I think they are mooning us!" Apparently, the bioluminescent rear ends of two fireflies (actually beetles) flush on their front window produce the mesmerizing twin lights emanating from the house across the way. Bioluminescence is, as its name implies, biologically produced light. It is produced through simple chemical reactions, by organisms as diverse as bacteria, protists, mushrooms, insects, worms, molluscs, corals, jellyfish, comb jellies, and fish. When they bioluminesce, these organisms make a range of colors. Larson's cartoon is effective even without knowing what color the mooning event was. If they were fireflies, the light was most likely yellow.

The reactions and molecules of different organisms are oftentimes quite different from each other. But all bioluminescence involves reactions of molecules. The molecule luciferin is critical in the reaction as a substrate that a second molecule called "luciferase" acts on. The kind of luciferase dictates what color of luminescence is emitted. When luciferase reacts with luciferin that has had an oxygen added to it (oxidized luciferin), the chemical oxyluciferin is produced and a photon of light is emitted. The photon shoots out at a specific wavelength and is then detected by any photoreceptors (in eyes or not, see chapter 3) that might be in the path of the photon. Beetles and other land invertebrates make yellow, green, or red luminescence. In contrast, bioluminescent marine creatures and ocean organisms make blue light, because it is transmitted more efficiently through water than the longer wavelength yellow and red.

Some organisms that luminesce do so because they have sequestered other organisms that do, or have ingested organisms where they can sequester the substrate and luciferase that produce the luminescence reaction. For instance, the luminescent bobtail squid sequesters cultures of a bacteria called *Vibrio fischeri* (also called *Aliivibrio fischeri*) in a small organ in its mantle, called a "light organ." The squid cultures the bacteria by diverting sugar and other nutrients to the light organ, and the bacteria in turn facilitate the luminescence reactions that emit light. The light emitted by the light organ camouflages or hides the squid by counter-illuminating the squid's mantle from below, matching the light coming into the top of the mantle.

Protists produce bioluminescence too. These species produce an amazing blue light when they come into contact with other organisms. The reaction is a survival response, as the blue flash of light acts as a sort of burglar alarm. Twenty-five years ago one of the authors of this book went midnight kayaking on Bioluminescent Bay in Vieques, Puerto Rico, with his daughters. You can's swim in the bay now as conservation measures for the organisms in it have been rightly instituted. Having adventurous offspring can be nerve-wracking, but their adventurousness on this trip created one of the most beautiful sights the author has ever seen. With no warning, his daughters jumped from their kayaks into the bay to take a swim. As they hit the water, they disrupted the billions of swimming dinoflagellates in the bay near them. This disruption triggered chemical reactions by the small organisms, which then produced dim blue light that outlined each of their swimming bodies to give an overall ethereal picture of two little blue angels floating in dark water.

Fluorescence is yet another way that light is manipulated by organisms. It is different from bioluminescence in that light is not directly produced by the molecules and a chemical reaction. Instead light interacts with an organism, is altered, and is then reflected as fluorescence. Fluorescence is based on the fact that organisms produce proteins called "fluorophores."

These fluorophores absorb visible light and create new light emitted at a different wavelength than what first hit the fluorophore. In essence a photon is absorbed by the fluorophores, and new photons with different, lower energy is emitted in a random direction. The wavelengths of the emitted light are quite different from the input light. Depending on the fluorescent molecule and the source of the fluorophore that is hit by the visible light, different colors will be emitted. This is why some organisms will emit green fluorescence, others red fluorescence, and still others blue fluorescence.

The range of fluorescence in ocean creatures is both dizzying and dazzling. Our colleagues David Gruber and Vincent Pieribone have spent most of their careers looking at the dynamics of fluorescence in marine ecologies. Gruber in particular started out interested in proteins, but he was quickly lured to the sea. Being an incredibly observant biologist led to his discovery of a green fluorescent eel that was found, against all odds, swimming in the night ocean. Reasoning that if there are fluorescing eels, then there are fluorescing anythings, Gruber set out about ten years ago to characterize this bizarre light phenomenon in a multitude of organisms and to try to get a glimpse of marine life that no one had previously seen. Since that fluorescing eel, he has been very busy. In a single expedition he found over two hundred new fluorescing marine organisms, including stingrays, bony fish, and sharks. To Gruber, who views the marine world through a yellow filter to enhance the fluorescence of the marine habitat, the diversity of life there is astonishing and looks completely different from what our trichromotic retinas can see. And because the yellow filter mimics what most fish lenses do, Gruber can view the ocean the same way most fish do.

Gruber's work demonstrates the grand diversity of fluorescent coloration in the ocean, but what about landlubbers? In fact, there are fluorescent terrestrial organisms, and they are somewhat diverse. Plant nectars fluoresce, as do specific areas of leaf surfaces and flowers of some plants. These fluorescence effects apparently serve as attractants for pollinators. Pitcher plants are exceptionally well studied in this respect and offer the strongest evidence for fluorescence as a visual attractor for insects and small mammals. Fruits of many plants can fluoresce, but this phenomenon has not been studied well enough to develop an explanation for its existence. Animal fluorescence is also widespread and well known in arthropods, especially in insects, scorpions, and arachnids. Fluorescence can be found in butterflies, beetles, cockroaches, grasshoppers, and dragonflies, but the exact role that fluorescence plays in the biology of these arthropods is not known. It could be involved in courtship, species recognition, or a wide variety of other evolutionary mechanisms, or it could simply be a by-product of other non-fluorescent pigments in organisms. Lizards and birds appear to have fluorescent capacities built into their bone tissues. For instance, in some species of gecko (family Gekkonidae) and chameleon (family Chamaeleonidae), bones fluoresce, and the emitted fluorescent light, is visible through the scales covering the bones. The blue range fluorescence of several species of chameleons in the genus *Calumna* has been characterized, and the role of this phenomenon in mating signals and sexual selection has been indicated by preliminary analysis of this group. In some bird species, their bills appear to fluoresce: in parrots, about 70 percent of the species surveyed so far have fluorescent plumages, and these are most likely involved in courtship displays.

Both bioluminescence and fluorescence are dynamic, protein-driven phenomena. Bioluminescence involves proteins that create substrates and enzymes to catalyze reactions to produce the release of photons. Fluorescence uses a pigment like chromophore to alter the wavelength of light.

Organisms that are iridescent form a third kind of natural way, without protein interactions, to change the properties of light and create natural colors and patterns. Because iridescence uses structures on organisms to implement the color change or production, it is also known as "structural coloration." Tom White, an expert in color patterns in nature, defines iridescence as "a change in hue with viewing or illumination geometry," a fancy way of saying that iridescence works by playing tricks on the eyes with angles. Illumination geometry is all around us.

Think of this. A rainbow forms by water droplets in a cloud, or from a garden sprinkler, and it is a beautiful sight. You just have to find the right position and figure out how to look at the right angles in relation to the sun to enjoy one of nature's most amazing light shows. The physics of the phenomenon is refraction, in this case the propagation of light in small spherical objects, the water droplets. Light at different wavelengths is propagated in different directions, and we see these specific wavelengths as colors enhanced against the general background of white scattered light.

A similar colorful effect can be seen in soap bubbles, which are thin sheets of water held stable by the surface tension at the interface between water and air. The soap helps generate such large, thin films of water. The thickness of the film varies from about one wavelength of visible light (about 1,000 nm) to about ten or twenty wavelengths of visible light. The light waves bounce forward and backward between the two surfaces, and some of the waves escape. Some light goes back, is reflected again, and joins with the earlier wave. This effect is called "interference." It happens for all wavelengths contained in the sunlight, and some of the wavelengths will cancel, some will enhance. The result is a selection of wavelengths that dominate over others, and this determines the color of the soap bubble viewed from this specific angle at any one point. Due to gravity, the thickness of the water film varies from the top of the bubble to its bottom, so the colors will gradually change from top to bottom. The colors change in real time, since the water slowly flows downward, and the film of water gets thinner overall. Such an effect of gradual change of color in nature is called *iridescence*. If much of the light remains white, it is called *pearlescence*, obviously named after the visual effect pearls display in certain kinds of light.

Watch this visual spectacle carefully next time you blow some soap bubbles. Alternatively, you could place a drop of oil on still water, whether in nature or in a small dish. The oil will spread out. The two liquids will not mix if you hold still. The oil will form a thin film of even thickness, getting thinner at the edge of the spot of oil. Again, the light is reflected and bounced back at the surfaces, water to oil and oil to air. Parts of the light are bouncing forward and backward, and the different waves interfere. This varies with the wavelength, and colorful effects occur.

Something related but quite different are the beautiful colors generated by some beetles and butterflies in their wings. This is a different form of coloration, due to the structure of the surface of the beetle's carapace and the butterfly's wings. The effect has been noted and understood in general terms since Robert Hooke published his book *Micrographia*, in 1665. Hooke noted that the iridescence of a peacock's feather was lost when it was plunged into water but reappeared when it was returned to the air. He reasoned that pigments could not be responsible for the color effect. It was later found that iridescence in the peacock feather is due to a complex pattern of crystals in the wing that are the size of visible light wavelengths. Currently, the same light effects are artificially made with photonic crystals.

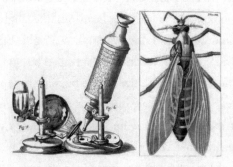

Figure 5.1. Hooke's microscope next to his drawing of a gnat. *Public domain.*

This coloration effect is created by very small structures that appear in periodic fashion on the surface of the wings of insects that are iridescent. These scales reflect light but also form small steps, of a height similar to the wavelength of visible light. Consequently, each step reflects a wave of light, and the waves reflected from two neighboring steps is shifted a fraction of a wavelength. Waves reflected from many steps can all propagate toward your eye, and the many partial waves interfere right at the entrance of your eye. Again, some wavelengths are suppressed due to the cancellation of waves, and some are enhanced. In physics we call this the diffraction of light with a grating. A systematic study shows that for a specific direction, many visible wavelengths lead to cancellation, while a few are strongly enhanced. For this reason, diffraction can form very strong effects and show very pure colors emanating from surfaces. The scales on the wing of an insect form such a grating. You can also see this effect from the gentle color changes, or iridescence, of the light reflected off a CD or DVD. The surfaces of these objects have gratings imbedded in them related to the purpose of the disc.

Polarization is another light effect that colors the world. This effect is caused by another wave property of light. Light as a wave propagates in specific directions. Let's think of a ray of light coming through a small hole in a curtain. The wave oscillates orthogonal, sideways, to this direction. Scientists call this a "transverse wave." There are two ways of being transverse. If you look at the ray of light head-on, like a light coming from a distant car, it could be up and down, or left and right. Light with waves only going up and down we call vertically polarized. The light with waves left and right we call horizontally polarized. It is possible to separate any light coming from the car headlight, say, into these two components.

The important thing is that it is impossible for these two components to interfere with each other. They stay separate but can be added to each other. The total light can have a clearly defined direction of polarization, vertically, horizontally, or, if both are present, at an angle between horizontal and vertical. This angle can be fixed, or it can rotate. The latter is called "circularly polarized," with the direction of polarization moving in clockwise direction coming toward us. This light is called "left-hand polarized." Any light that is in anti-clockwise direction is called "right-hand polarized."

While humans are continually inventing new ways to bend light and new materials to manipulate light, organisms have evolved to take advantage of all these light phenomena. Iridescence, pearlescence, and polarity are all light phenomena that organisms have exploited in the evolutionary process. As a result of a detailed examination of iridescence, Thomas E. White points out that colors in nature that work adaptively do so because the signals they give are constant and stable when "painted" onto a somewhat colorful chaos.

With respect to vision, the brightness of something will vary over different times of the day, in different kinds of weather, and in different ecologies. Brightness can vary over orders of magnitude depending on context, hence it is not such a good communication signal with respect to light. There are some uses where brightness can be useful, usually in warning contexts.

Since colors are stable, they are much better at communicating important evolutionarily derived signals, such as in the mating and courtship process. White states that signals conveyed via hue are so clean and reliable that they can be used to establish identity, category, and quality of organisms in a stable adaptive evolutionary system. Iridescence throws a wrench into the picture, as it mucks with the stability of hue of many objects in nature that depend on color cues for mating or other behaviors. While White sees iridescence as a major factor in generating color signals in nature, he points out a paradox: How can a mechanism that enhances the instability of color be effective at exchanging information in an adaptive context? Understanding the evolutionary nuances of iridescence is indeed important, and work is needed to, as White says, "suggest a pathway to understanding the function of this optical curiosity, and with it an entire dimension of biological diversity."

Lewontin and Gould Spoil the Party

In the last two chapters we have discussed the wide range of color and color variation in nature and its relationship to natural selection. While phenomena like Batesian mimicry, camouflage, sexual selection, and the like can be pitched as great examples of natural selection, we would feel remiss if we did not introduce a strong caveat here. And while we have leaned heavily on natural selection as a means to explain the amazing diversity, there might be other explanations. As we have pointed out, much of the research program after Darwin and Wallace's breakthrough focused on characterizing natural selection and adaptation in nature. The formulation of theoretical population genetics and evolutionary theory and the emergence of the Modern Synthesis by the 1950s cemented the search for adaptation as the prevailing research paradign. Stephen J. Gould and Richard Lewontin, Harvard colleagues and partners in crime in shaking things up in biology, grew tired of the prevailing paradigm of evolutionary research mostly because they felt researchers were looking for adaptation much too blindly and uncritically. Both are well known for their research contributions to evolutionary biology, and both were expert communicators of science. Both authors have always been entertaining, but in completely different ways. Lewontin, more famous for his brusqueness, once ended a paper with the phrase "Tough luck" in reference to his belief (well founded) that the genetic basis of many human behaviors would never be worked out. To him they were so complex and emergent that biological explanations for human behaviors (like preferences for religion, politics, and other socially embedded behaviors) were simply not possible. This attitude grated on a large number of researchers who had based their careers on dissecting human behaviors in the realm of sociobiology and evolutionary psychology. Gould, on the other hand, was famous for his capacity to spruce up explanations of the evolutionary process with examples from literature and other walks of life (he was an avid baseball fan and a follower of Mickey Mouse).

Together they published a beautiful paper in 1979 called "The Spandrels of San Marco and the Panglossian Paradigm." The mention of St. Mark's Cathedral and Dr. Pangloss in the title are indicative of their creativity in communicating science. Dr. Pangloss refers to a character in Voltaire's play *Candide* who held eternal optimism about finding reasons to explain away almost any observed phenomenon (we have noses, therefore we wear eyeglasses). Gould and Lewontin drew a comparison between Dr. Pangloss and researchers in the 1970s who were focusing on adaptation as explanations for almost everything; hence the Panglossian Paradigm. The reference to St. Mark's is much more pointed though. They describe the ceiling of St. Mark's Cathedral as the perfect canvas for many of the biblical stories that were painted there. As they suggest, one could almost claim that the spandrels, which you can see when you look up at the ceiling, are there to frame the biblical scenes. But of course, as the authors continue, the spandrels are there to hold up the four large domes that provide the wondrously beautiful space inside the cathedral. The point they were trying to make is that scientists can make observations and come up with explanations, but if they don't dig deep enough, they might fall into a Panglossian trap. The rest of this seminal paper suggests that adaptation had become too much the focus of evolutionary biology up to the 1970s, and researchers had found themselves in an adaptationist rut, where they could see adaptation everywhere. Maybe we need to step back and take a closer look at color in this way too.

Part of the reason that we focus on the genetic basis of many of these traits is that once this aspect of a color trait is worked out, the precise nature of natural selection can be worked out. Otherwise it is incredibly difficult and tedious to prove the link between natural selection and the patterns of color variation we see on our planet. The genetic information allows us to address directly where natural selection actually works, and that is at the level of the genome. We now know the genetic basis of melanism in this system. As with many of the genetic mechanisms we have discussed in this and preceding chapters, it is not a change in a structural gene but rather a regulatory effect. The cortex gene makes a transcript that is involved in the melanistic transition, and this gene in the melanic form is interrupted by a transposable element called *carbonaria*. The interruption of the cortex gene by this transposable element affects the wing development, which is in turn involved in the melanism. This is not what was expected at all, and indeed it throws a monkey wrench into understanding this iconic system.

While the review by Cuthill and colleagues mentioned earlier in this chapter has set the color research paradigm and is a great extension to the past century of research on color, we can't forget that there is a Panglossian pit out there. Prior to modern approaches to color, evolution research was very localized to single researchers or very small collaborative approaches.

The extension of many of these classical color systems into genomics and other modern technologies can only keep us from falling into that pit. Cuthill and colleagues recognize this too and point out that the collaborative approach to color in nature is a good way to avoid the pit. They state, "We are on the threshold of a new era of color science, and the interdisciplinary nature of this collaborative enterprise holds enormous promise."

6

The Colors of History and Culture

A t the risk of requesting that you the reader do something very cliché-ish, we ask you to imagine waking up to a world without color. You can't see colors, neither can others around you, if they are humans. You see everything in black and white, like in an old movie. In this world, you and all of the humans around you are reacting to what you are seeing in a very different way than you do with color vision today. How do we know this? It turns out that color influences nearly every aspect of our existence minute by minute, second by second. As humans we get cues from color from every angle and perspective. This is because color has been ingrained in our collective psyches as a series of cues important to our survival in groups. Groups of people develop cultural practices and standards and these have molded most of how we react to colors in a cultural context.

Ancestral Colors

What color means to people varies broadly across cultures. The approach to understanding color in a cultural context starts with our ancestors and how they perceived color, what color meant to them, and how they might have used color. As we have pointed out, our ancestors have a variable history with respect to color, depending on how far back in time we want to go. If we are talking about our primate ancestors, then we are talking about a specific history, a unique one with respect to other mammals. Evolutionary reconstruction is a simple approach that looks at the distribution of traits in living organisms, and it makes clear that some of our primate ancestors had a trichromatic way of viewing the world.

Let's try to reconstruct a trait in primates. We would look at as many of the primate species as we could for the trait. We would also look at the same trait in organisms outside of primates but still reasonably closely related. By looking at the phylogenetic distribution of the trait in primates and in other organisms, we can tell whether or not the ancestor of primates had a particular trait. Then, using the same approach, we can trace the trait across the whole of primates.

Bony fish, reptiles, and birds have the capacity for tetrachromatic vision. That is, they all have four kinds of cone opsins. Remember, there are three major cone opsins in us humans (making most of us trichromatic): SWS (of which we have two, SWS1 and SWS2), LWS, and MWS. It is safe to say that fish are diverse with respect to their cone opsins, and the common ancestor of fish more than likely had at least four kinds of opsins, as have the bird and reptile common ancestor. Since mammals are related to the bird/reptile group, it is safe to assume that the common ancestor of vertebrates had at least four cone opsins. Next let's look at marsupials and a few of their placental mammal close relatives. Because marsupials are the closest relative of placental mammals, the number of opsins in their genomes is important information for inferring placental mammal opsin gene maekup. Marsupials have only two cone opsins (making them mostly dichromatic, but there are some trichromatic marsupials), while most of the mammal groups have the same two cone opsins also. What this means is that in the common ancestor of mammals (placental mammals plus marsupials), there were only two cone opsins. This common ancestor somehow lost two of the opsins as it diverged from its common ancestor with reptiles and birds, who were tetrachromatic.

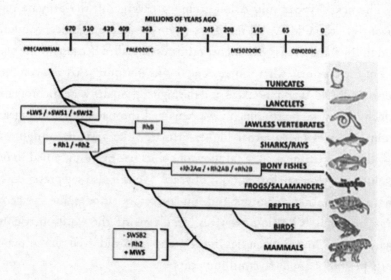

Figure 6.1. Phylogenetic tree showing where in the chordate tree of life the various opsins have arisen. The four opsins and two rhodopsins that are present in our species are shown in bold. Other lineages like the bony fishes have accrued different kinds of rhodopsins. The + and − signs indicate gain and loss, respectively. *Drawing by Rob DeSalle.*

Next we need to look at the different kinds of primates out there. The way that primates diverged has been a major topic for anthropologists and biologists for a couple of centuries, so we know a lot about the patterns of divergence in this group. Lorises plus lemurs is the first major group to break off of from the primate ancestor (which had to have two cone opsins). Next comes tarsiers, followed by New World monkeys (monkeys like howler monkeys and tamarins). These organisms exhibit the phenomenon polymorphic trichromacy (or the other side of the coin, polymorphic dichromacy). This phenomenon is where both dichromatic species and trichromatic species exist in the groups, meaning that dichromacy is the ancestral state for all of these kinds of primates and trichromacy evolves independently in the groups to produce the polymorphic trichromacy.

The next African and Asian group to diverge on the primate phylogeny are the Old World monkeys (primates like macaques, baboons, and the like) followed by the higher primates (so-called because we are one). All of these primates have three cone opsins, meaning that the common ancestor of Old World monkeys and the higher primates were trichromatic, which is how we pretty much have remained throughout our divergence from ancestors of the other higher primates like gibbons, rangutans, gorillas, and chimps. So, a rainbow of colors has been ingrained in our evolutionary lineage for millions of years. We can also say pretty surely that our hominid ancestors and our ancestors throughout the genus *Homo*, as well as its close relatives, were viewing the world in trichromatic color. Dozens of species in the group hominid lived on our planet for a period of about five million years.

These species are all either in the genus *Homo* or in very closely related genera, starting with *Sahelanthropus* and *Orrorin* and leading to the *Australopithecus Paranthropus*. Evolutionary reconstruction suggests that these species, including Lucy (the prime example of *Australopithecus afarensis*), more than likely saw the world in trichromatic hues. Within our genus, which began its divergence about two million years ago, there are several species. Indeed, at any given time up until about thirty thousand years ago, more than one species in the genus *Homo* lived on the planet, with up to four or five species existing at any given time. Our understanding of the divergence of our closest extinct relatives is complex because of the paucity and relative incompleteness of fossils of these species. (Although there are some beautifully preserved *Homo* fossils, they are rare.)

One factor that has increased our understanding of the evolutionary history of these species is researchers' ability to sequence their genomes. The heck, you say! But yes, by making DNA from an ancient tooth or a small fragment of ancient bone, researchers like Svante Paabo and his colleagues in Germany have sequenced the genomes of dozens of ancient *Homo* specimens. This herculean task (it is a very difficult and highly controlled process because of contamination of the DNA sequences by bacterial and modern human sequences) can give researchers genome level information for these long-dead organisms. Currently the oldest part of a genome sequenced is from very old *Homo sapiens* at about 300,000 years old.

The history of these genus *Homo* species is complex, because it appears they were interbreeding with each other after their divergence from each other. *Homo neanderthalensis* (commonly known as *Neanderthals*), a robust species in the genus *Homo*, appears to be our closest extinct relative, but there is also a species (yet to be named based on fossil fragments) called "Denisovans" (because of their discovery at the Denisova Cave in Central Asia), that coexisted with *H. sapiens* and Neanderthals. Another recently discovered species, the Hobbit from Southeast Asia, is also in the game, but it is probably less related to Neanderthals, Denisovans, and us. There are even some "ghost" species of *Homo* that researchers hypothesize had to exist because of patterns in these ancient genomes, but we haven't found them yet. Jerry Springer would be proud of this work, as it is kind of like the ultimate parental testing system, only we don't yet know who the missing parent is.

We know with impunity that Neanderthals and Denisovans saw the world trichromatically, because researchers have generated their genome sequences. This research shows that Neanderthals and Denisovans had a full complement of trichromatic opsins in their genomes, and thus had the capacity to see color just the same as modern *H. sapiens*.

We even have some clues about how much color played a role in the cultural development of hominids, as there is evidence for the use of colors in Neanderthal settlements. They apparently used iron oxides and manganese oxides as pigments for decoration and doctored up marine shells for some as yet not understood purpose. Manganese oxide, hematite, and pyrite are particularly interesting, because they produce both red and black pigments, which probably were used to adorn the skin in some way. Recent discoveries (about fifty thousand years old) at Cueva de los Aviones and Cueva Anton have uncovered further evidence of color use on marine shells and have given science an unprecedented idea of how Neanderthals used color. Yellow, orange, red, and black seem to be the preferred colors of Neanderthals, and they also used crosshatching on bones and eagle talons as decorations. In addition, they used bird feathers, which were no doubt of different colors. While it is difficult to pin down why they were using these pigments, it is clear they were being used at the same time that our species was developing a complex utilization of color. It indicates the emergence of complex behavior for Neanderthals; behavior at least as complex as their temporal con-generics (our species) alive at the time. Neanderthals went extinct about 25,000 years ago, so Neanderthals probably used color for a short period of time. While, as we will soon see, our species probably started the use of pigments before that, it doesn't mean that Neanderthals weren't using pigments all along.

Table 6.1. Archaeological discoveries relevant to color use over 300,000 years of *H. sapiens* existence. From Colagè and d'Errico, 2018.

Items with potential for color use	Age (1,000 years)	Continentwith first use
Painted images	45	All
Rock art	60	Europe
Beads	135	Africa
Pigment processing	200–300	Africa
Colored objects	100	Near East

Neanderthals and *H. sapiens* split into separate lineages between 700,000 and 500,000 years ago. For a few hundred thousand years our species existed as an independent lineage known as archaic *sapiens*. It wasn't until between 200,000 and 100,000 years ago that the modern form of *sapiens* appears. Once that happened, we started to move all over the globe, coming into contact with Neanderthals, Denisovans, and whatever other species of *Homo* were out there. Sporadic interbreeding amongst all of these lineages occurred up until the extinction of Neanderthals and Denisovans. There is strong evidence from *H. sapiens* archaeological sites that pigments were processed from ores as early as 300,000 years ago (Table 6.1). But more precise understanding of the use of color comes from several archaeological sites. The discovery of the archaeological site at Pinnacle Point on the south coast of South Africa lends strong evidence that our species' use of pigments dates back to before 160,000 years ago. This important site yielded artifacts indicating exploitation of the marine environment for food and sustenance and a shift in diet in *sapiens* settlements. With those discoveries comes the possible use of pigments in everyday life. Other sites essential in the interpretation of pigments as important aspects of symbolic expression in our species are listed in Table 6.2. Another site at Qafzeh Cave in Israel also offers important insight into the use of color by our species. This cave, with artifacts dated as far back as 92,000 years, is a treasure trove of information on *sapiens'* use of ochre. We have mentioned this pigment product above but note that it is obtained from rocks rich in ferric oxide, clay, and sand. The pigment was obtained by fracturing the rocks and pulverizing it with hard objects. Researchers infer that these rocks and the ochre isolated from them were part of a complex behavioral system that was associated with symbolic logic.

It is clear that our species has had a long and close relationship with colors, and that colors more than likely had a lot to do with our ancestors' realization and application of symbolic logic. According to our colleague Ian Tattersall, this realization and use of symbolism is at the heart of human consciousness and had everything to do with how we as humans became the complex organisms that we are today (and were even back then). For tens of thousands of years, our species has been extracting, grinding, mixing, experimenting with color as a means to communicate something no other species—except, perhaps, for our close sister species Neanderthals—could or even wanted to express. There is a paradox here though. As we can see from Tables 6.1 and 6.2, pigment use goes all the way back to around 300,000 years ago. However, our species experimentation with and settling on symbolism is widely thought to go back to artifacts found in the Blombos Cave in South Africa, dated to a little over 100,000 years ago. What this means is that the visual draw of coloration from pigments was strong in early *H. sapiens*.

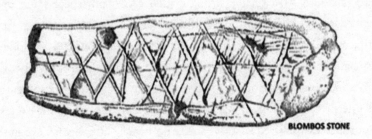

BLOMBOS STONE

Figure 6.2. The Blombos stone. *Drawing by Pat Wynne, included with permission.*

It's almost as if these early *sapiens* couldn't help themselves from using color because their eyes were so keyed into visual cues and visual stimulus. The colors they used and the processes they developed to make pigments in essence preadapted them to using color to expand symbolism and to hone the symbolism to such a degree that our unique consciousness emerged. We are not saying that it is color only that did this, but color is more than likely an important cog in the development of this consciousness about 100,000 years ago. At that point all hell broke loose, because it is apparent from archaeological sites caked with ochre on the floors of caves, sometimes inches thick, that it was in wide use, more than likely as part of cave art and body/face painting, as well as coloration of animal skins, shells, and other artifacts. All of this use of pigments was worldwide and extreme and included *sapiens* settlements in Europe, Asia, Africa, and Australia (*sapiens* only migrated to the Western Hemisphere seventeen thousand or so years ago). Extensive documentation of this prehistoric use of color comes from the proliferation of cave art and cave painting so evident from discoveries dated to only twenty thousand or so years ago. At about ten thousand years ago, our species made a huge cultural, social shift in the way they lived, going from vagile, semi-solitary hunter-gatherer societies to sedentary cultures living in groups. Color remained an important part of these early modern cultures' everyday existence. Just as we will soon see that our use of color today permeates every aspect of our day-to-day lives, so too these members of our species living tens of thousands of years ago immersed themselves in a sea of symbolic coloration.

Table 6.2. Key archaeological sites with ochre discoveries from Terlach, 2018.

Site	Items	Age
Olorgesailie, Africa	two pieces of intentionally shaped ochre	307,000
Blombos Cave, Africa	engraved piece of ochre	100,000
Northern Cape Africa	ochre fragments	500,000
Twin Rivers, Africa	pieces of ochre stained quartzite cobble	266,000
Porc-Epic, Africa	largest collection of ochre pieces	40,000
Maastricht-Belvédère Eu	ochre fragments	250,000
Rose Cottage Cave Africa	ochre collection	0–96,000
Madjedbebe, Australia	ochre pieces	65,000

Cultural Coloring Book

Once modern cultures began to develop, most bets about natural selection were off. But why? When distinct cultures are involved, different kinds of pressures influence the evolution of traits, preferences, behaviors, and other emergent properties of human populations. So far, we have taken a very genome-based adaptationist approach to understanding diversity on our planet (but see the end of chapter 5). There are other ways that behaviors, social preferences, and cultural traits can evolve in populations.

Evolutionary biologist Mark Pagel once said, "A camel would make a poor musk ox and a penguin a poor monkey." While this statement might seem out of left field at first, Pagel was trying to explain how genetics and the genomes of organisms have been molded by natural selection to attain the phenotypes they exhibit. The camel and the penguin work as the organisms they have evolved to be and would be poor substitutes for other organisms. Nor would the monkey or the musk ox be able to fit into the evolutionary slot that other organisms occupy. Hence a penguin no doubt would make a poor monkey. No organism on the planet could switch gears enough to occupy the evolutionary place of another—until, as Pagel points out, our species came along. For much of our existence as a species we couldn't switch slots either, until we developed the capacity to communicate symbolically in a cultural context. This is because of a mechanism that we obtained and what Pagel describes as cultural survival vehicles (CSVs). CSVs are at the heart of cultural evolution, a novel way that we humans have evolved over the past tens of thousands of years. Cultural evolution and our extensive use of CSVs are not like organismal evolution by natural selection, mostly because the force of natural selection is missing. While natural selection is the vehicle by which organismal evolution is molded, social learning is the vehicle by which cultural evolution works, and CSVs are the things learned and passed on. The outcomes of cultural evolution can be adaptationally illogical solutions to challenges associated with human cultural existence.

Cultural evolution is amazingly rapid in implementing change in human societies. Pagel points to three distinct things that it accomplishes: it carves up landscapes mostly along linguistic lines, it restricts the flow of genes in space, and it slows the flow of information from outside. Cultural evolution does all of these very rapidly, in just a number of human generations.

In addition, note that each of the three aspects of cultural evolution that Pagel cites tends to erect barriers, which explains the vast differences in the cultural behavior of people across the globe. For color vision, cultural evolution and the involvement of color in many CSVs mean that color is a fascinating topic with surprises at every turn.

It is safe to say that the interpretation of color differs among cultures. But are colors perceived differently by different cultures? Imagine the difficulty of addressing this question. The answer to the question is tied directly to language usage. There are about 6,500 existing languages on the planet, with many going extinct (cease to be spoken) each year. This means that for the three prime colors—red, yellow, and blue—there are at least 19,500 words that can be used to name them. Some languages overlap in the names they give colors, but for the most part each language has come up with their own words, their own vocabulary for the colors they see. When all of the variation in hues that cultures recognize are added to this twenty thousand or so words for color, it creates a very complex and hard-to-solve problem. More than likely no two humans (except for perhaps identical twins) perceive colors in the same way. This is partially due to the individual variation of our vision sensors, with color blindness being the extreme case, and partially due to learning within a group of humans, within a society.

There is indeed a fairly high degree of variation at the genetic and DNA sequence level in the genomes of humans for opsin genes. Some of the variation is silent (DNA sequence change that does not alter the amino acid sequence of a protein) while other DNA sequence variants code for changes in the visual pigment proteins that alter color vision in some way. However, even the most common color vision problem occurs in only 8 percent or so of humans (see Table 6.3), and we need to remember that the frequency of these variants varies from population to population. Since there are three major kinds of opsin pigments in the average human genome (LW, MW, and two SWs), the common state can be considered to be trichromacy.

However, there are many ways this state can be altered. Most readers will be familiar with mutations at the DNA sequence level, and such alterations in the primary sequence of genes in regions of the gene that codes for proteins could indeed result in changes in the structure of the three kinds of opsin pigments. The online database for inheritance in human populations, the Online Mendelian Inheritance in Man (OMIM), lists hundreds of nucleotide changes that have occurred in opsin genes in human populations. Another and more common mechanism for altering how humans perceive color is to lose or gain whole opsin pigment genes. Because the red and green opsin genes are adjacent to each other in the genome, a mechanism called "unequal crossing over" in the germ cells can sometimes lead to the loss (or gain) of opsin coding regions, creating dichromatic color vision. And, of course, there is the simple possibility that any of the three kinds of opsin genes could be lost from a genome or completely inactivated. It turns out that human populations have almost all of these kinds of changes in opsin genes that cause color vision anomalies.

There is a whole classification of such anomalies: protanopia, deuteranopia, and tritanopia are color vision anomalies in humans caused by the loss of one or more opsin genes. Protanopes cannot distinguish between red and green colors. Protanopia is caused by the loss of the LW opsin. Remember that this opsin and the SW opsins are important in the opponency of red and green light (see chapter 3), hence the detection and perception of red and green light. Deuteranopes are missing MW opsins. Like protanopes, they cannot distinguish reddish and greenish colors, because missing the MW opsin means opponency doesn't work for these individuals either. Both protanopes and deuteranopes are said to be red-green color blind, but they can be distinguished from each other by the way they process relative luminosity, or the intensity of light. Protanopia shifts the luminosity function toward shorter wavelengths, so the individual with protanopia has trouble seeing any light in the long (red) end of the range that humans process, while deuteranopes can't perceive green. This leads to slightly different views and perceptions of color by a protanope and a deuteranope. Finally, tritanopes are missing SW opsins. They cannot detect blue light and confuse yellowish and bluish hues.

Trichromats—individuals with all three kinds of opsins and hence all three kinds of cone cells—can also have low-functioning opsin genes or can simply develop their cone cells with a low number of opsin proteins in them. Weak LW opsin proteins or low numbers of the protein results in protanomaly. Weak MW opsin gene function or low numbers of this protein in cone cells leads to deuteranomaly, and weak SW opsin function leads to tritanomaly. In addition, to these cases where a single opsin type is mutated, leaky, or missing, some people can have two of the three kinds of opsins missing or non-functioning. Such individuals are called "monochromats," and they have only a single kind of cone cell in their retinas. Table 6.3 shows the percentages of modern human populations with the various kinds of dichromatism and monochromatism.

Table 6.3. Hunt (1991a) percentages of color vision anomalies

Type	Male (%)	Female (%)
Protanopia	1.0	0.02
Deuteranopia	1.1	0.01
Tritanopia	0.002	0.001
Cone monochromatism	~0	~0
Rod monochromatism	0.003	0.002
Protanomaly	1.0	0.02
Deuteranomaly	4.9	0.38
Tritanomaly	~0	~0
Total	8.0	0.4

As Table 6.3 shows, only about 8 percent of the general human population is impacted by these color vision anomalies, mostly males. This means that most human populations are trichromatic, and the vast majority deal with colors in a trichromatic context.

Are colors perceived differently by different cultures? Imagine the difficulty of addressing this question. The answer is tied directly to language usage though. Two anecdotal cases of language-based differences in color usage cited by Arya D. McCarthy and David Yarowsky and colleagues will suffice to make the point that perhaps not all languages have the same number of names for colors. In Korean, a single word (*pureu-n,* which means both grass and sky) can often correspond to very different colors. Another example is the English word *blue.* Blue has many cannotations and substitutes in English—aquamarine, navy blue, etc. But simply and generally when we say blue we recognize the whole range of blues. But Russian speakers split the generic blue into *goluboy* (light blue) and *siniy* (a darker blue).

The development of language and vocabulary around color becomes a complex but very important nut to crack. Paul Kay and Brent Berlin addressed this problem in a study in 1969 where they claimed that there "exist universal cross-linguistic constraints on color naming, and that basic color terminology systems tend to develop in a partially fixed order." Wow! These are amazing conclusions and suggest an inborn capacity for the language of color. But are these conclusions real? After all, they only used twenty different languages to come to their conclusions, and this was only one of several criticisms of the study. What was needed was a larger sample of languages from multiple individuals from each language and more detailed analysis of the usage of language in the context of color.

Kay and Berlin decided they would address these issues empirically in 1976 when the World Color Survey (WCS) was first accomplished. This survey broadened the number of languages used and established a color palette for how color would be observed and quantified for linguistic analyses. This test palette consisted of a forty-by-eight matrix of different colors, plus blocks of colors from the spectrum. Hence there were about 320 chips of distinct colors tested with each person. Each subject was asked to name the color in the corresponding chips. The linguistic patterns from this analysis were obtained for 110 languages, all of which are unwritten languages. The reason for using unwritten languages is that they are presumed to represent very basic language families and hence are a more fundamental source of information for addressing the issues involved in tracing language usage in the context of color. The methodology of the analysis was to take the answers of the people being tested and present them as a matrix of responses. The matrices could then be easily analyzed mathematically for patterns. Figure 6.3 shows an example of the transformation of the responses of an individual from a specific language into a code that can be analyzed mathematically.

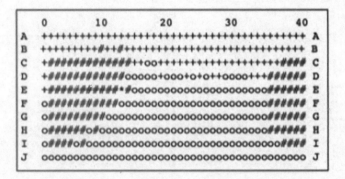

Figure 6.3. The Kay and Berlin matrix for an individual speaking model Wobe'. The symbols are: o = kpe' or black/green/blue; + = pluu- or white; # = sain' or red/yellow. *Drawing by Rob DeSalle.*

The symbols in the table represent specific responses of one of the Wobe'-speaking individuals. For instance, the # symbol represents a specific linguistic response (*sain'*) for that color in the chip. It turns out that for this sample, over sixty chips were given the same name corresponding to the o (*kpe'*) symbol. Multiple individuals using the same language were tested, and an aggregate measure of agreement of color terms can be computed. Each specific color (or color combination) for each language can then be represented by an aggregate color term map, as shown in Figure 6.4. The figures represent the color term maps for the Wobe' language for two color terms, *kpe'* (black/green/blue, on the left) and *sain'* (red/yellow, on the right). Once color term maps are constructed for different languages, they can be compared across the languages. The color term maps take on an iconic appearance of their own (see Figure 6.4), and these appearances can be quantified and used as data in further analyses.

The color term maps are compared to each other using clustering approaches to determine the relatedness of color representation of the various languages in the study. Once the patterns and metrics of a large number of languages are curated, different hypotheses about the development of color naming for humans can be tested.

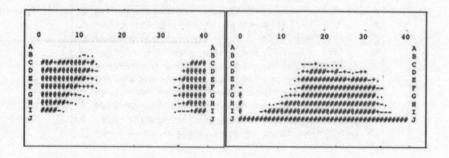

Figure 6.4 Color term map for the word kpe' or black/green/blue for Wobe' on the left compared to the color term map for sain' or red/yellow for Wobe' on the right. Spaces in figures are aspects that could not be scored. *Drawing by Rob DeSalle.*

Cluster analysis simply uses two-dimensional or three-dimensional graphs, where different languages or words for different colors are plotted based on their color term maps. Languages with similar color term maps will cluster together in 2-D or 3-D space. An even more precise approach is to combine each color's (red, green, yellow, green, blue) color term maps on the same graph. When this is accomplished, discrete clusters of different languages are observed for each color, indicating that the color naming process in these different languages all have a similar basis. Hence their statement that "universal cross-linguistic constraints on color naming" exist. They also used the data to reconstruct the progression of color naming from more basic and combinatorial color naming (white+yellow+red vs black+green+blue) to less inclusive naming (only red, only yellow, and so on) and discovered universals in how different languages have developed color naming. While many researchers in linguistics accept these conclusions, there are some who still question its universality. However, recent analyses using the WCS with a larger sample of languages shore up the universality idea. Arya D. McCarthy and her colleagues examined color usage in 2,491 languages (both written and unwritten) taken from online resources on language. Their computational model of color usage, while obtained by a very different method than Berlin and Kay's original study, shows a strong correlation with the 1969 study, leading McCarthy and her colleagues to support the original claims. Delwin T. Lindsey and Angela M. Brown used the WCS and applied different computational methods than Berlin and Kay's original ones. They examined color naming motifs, which they found to be distributed worldwide, and concluded that such "distribution of the motifs and the co-occurrence of multiple motifs within languages suggest that universal processes control the naming of colors." They further conclude that there are eleven basic color naming motifs that all languages use, suggesting a striking similarity in what humans across the globe are seeing with respect to color.

The broad variation of color perception in humans that leads to varied uses of color across human cultures tangles with the universality of color word usage in languages. This tug-of-war has led to distinct differences in how color is used and what colors mean in different cultures. While color language usage has some universals, color symbolism does not. A preference for a particular color is like one of Pagel's CSVs. It can easily sweep through a population as a result of cultural evolution and get set as part of the population's cultural milieu. There may be some cross-cultural linguistic universals when it comes to color, but the color CSVs that occur are fickle, and this leads to some strange apposed preferences in human populations and cultures. Take the color red, for example. Royalty, strength, life, blood, wealth, power are just a few of the things that red has stood for throughout history. But it isn't as simple as that. As many readers will know, red can mean different things within a single culture.

Pinkie, Blue Boy, and Pink Boy

The use of red in clothing often means power, or "look at me—I stand out." Look no further than the 2016 presidential race in the United States, where one candidate is famous for wearing an overly long red tie and the other a red power suit. The stylist Bill Blass once said, "When in doubt wear red," probably because if you aren't interested in making a specific statement with color, the best thing to do is make a power statement. Since red can also mean "look at me," it is one of those colors that has a dual meaning in Western cultures. Often it is considered garish or less than honorable. Just think red-light district, or Sting's lyrics from "Roxanne": "You don't have to put on your red light." And while there were some pretty tacky tuxedos in the 1960s and 1970s, the color red was so tacky that very few males wore red ones to the prom. Today the staid color black is the norm for tuxedos. Tacky or dishonorable uses of red directly contrast with its use as a symbol of power. Red conveys emotions too, like passion and romance, anger, dangerous feelings, and volatility.

Colors are often used to establish group membership and form bonds. While there are many examples in American sports, the color red is used throughout the globe for sports teams as a way to make group (be it team or fan) divisions easily recognizable. But it's not just sports where the color red and other colors are used to engender group identity. In the United States, political parties are identified by red and blue—so much so that a state is designated as a red state or a blue state depending on whether the state leans toward conservativism (red) or liberalism (blue). In Australia, it is reversed, as the Liberal Party (which leans more to the conservative right) is represented by blue; the Labor Party (which is liberal leaning) is represented by red; and of course, the Green Party is represented by green. In most European countries the left-leaning parties are represented by red. While these color preferences for political parties are rather new, using color to identify political philosophies is older. The red flags of the People's Republic of China and the now dissolved USSR are good examples of red representing political movements in these countries. The red in the flag for the USSR goes back to the French Revolution, where a red flag, or Red Banner, was oppositional to the upper class and represented protest. Later the Paris Commune of 1871 adopted a red flag. The use of red in both earlier movements were adopted by the USSR as it was being formed as the color of the national flag. The red in the Chinese flag has a much deeper history though. Confucianism used red as a symbol of luck, happiness, and joy, and this meshed nicely with the Marxist leanings of the Cultural Revolution in China. It is almost as if the red in the Chinese flag was preadapted to represent Maoist thinking. Red pops up in Mao Tse-tung's *Little Red Book*, in the de facto anthem of the Chinese republic ("The *East* Is *Red*") and the famous chime of the Chinese people that Mao is "the red sun in our hearts." Other non-Communist cultures developed an aversion to red. Not wanting to be associated with Marxist, Maoist, or Communist thinking, they chose yellow as the go-to color for protest.

Red is also used ceremonially, including in weddings and religious rites. In India brides wear red saris, which represent fertility and prosperity for the Hindu women who are marrying. Red also has the traditional meaning of love in Hindu cultures, hence the use of red in a wedding rite. Many Asian cultures have adopted red as a sign of auspiciousness, and red in the Hindu wedding rite appears not only in the bride's dress but also in other objects used in the rite representing good luck and providence. Contrast this with the now popular use of white dresses for the bride in Western wedding rites. In 1840 Queen Victoria wed Prince Albert in a white dress, and this started a trend in the West of white as a wedding dress color. White in this context also meant wealth because of the tradition of wearing the dress only once. Prior to the tradition, brides wore their favorite dress at weddings, but if they could marry in white and then store the dress, this was a sign of wealth. Red also was developed as a very important color in the Catholic religion, as it represented the blood of Christ. Red was also the color of the cardinal's robe and signified the willingness of these religious leaders to shed their blood for Christ.

Red was a difficult color to create before modern times because there was no readily available red dye for clothing. While red ochre could easily be created from specific kinds of earth, it was a dull red and not a very good dye for clothes. As the New World was colonized, dyes that fit the bill for various colors were discovered and brought back to Europe. In 1523 the Spanish explorers returned with a brilliant red dye made from cochineal, which quickly became valuable and rare in Europe. It comes from the insects (*Dactylopius coccus*) that live on cacti in Central and South America and was discovered by the Native Americans from these regions over two thousand years ago. People from these regions cultivate the cochineal, dry them, and crush them to produce the brilliant red dye. They kept the process a secret as well as they could, so the dye became rare and coveted by Europeans. The rarity of these red dyes allowed only those of high status to use them. It is not surprising that Europeans developed a penchant for the color red. Among the Inca, red conveyed a high status, as only the king was allowed to wear red clothing that was dyed using cochineal. The Aztec rulers also recognized the sign of high status from the color red in general, and they would demand tribute in the form of cochineal dye.

Even if we go a little off red to pink, the cultural ramifications are still significant. In the 1700s, three portraits—*The Blue Boy*, *Pinkie*, and *The Pink Boy*—foreshadowed the use of pink and blue to indicate masculinity or femininity. *Blue Boy*, painted in 1770 by Thomas Gainsborough, depicts a young man decked out in blue pants, jacket, socks, and shoes. He maintains a very masculine stance, with one hand on his hip holding a blue cape. In his dangling hand he holds a dark blue hat. His hair is a bit tussled, and his gaze is very boyish, as anyone with a tweenish son can attest. *Pinkie* was painted in 1794 by Thomas Lawrence. It shows a young girl in an ankle-length pink dress, with a dark pink sash across her waist. The dress is flowing in the wind. She wears a pink bonnet with the ribbons also flowing in the wind. Pinkie has one hand behind her back and the other crossing her body just below her neck in a quite graceful pose. Her face is blushed pink throughout, and while Blue Boy has some pink in his cheeks, the contrast between the two faces is striking. Twelve years after *Blue Boy*, Gainsborough painted *Pink Boy*. This boy strikes a masculine pose similar to that struck by Blue Boy. Pink Boy holds his hat in his left hand, his right hand free, his face blushed as pink as the suit he is wearing. Fast-forward to the early 1900s, and it would not be unusual to dress a male baby in pink or blue in the United States and Europe. But as the 20th century progressed, pink became the color for girls and blue the color for boys as a result of targeted marketing of wholesalers. This artistic pink and blue dichotomy is what we would call a "weak" or "neutral" CSV. Preference for blue as masculine and pink as feminine wavers back and forth even across generations. As we implied earlier, parents were confused in the early 1900s as to whether pink was a good color for their male offspring. In the 1960s and 1970s, pink was entrenched as a feminine color.

Words get created all the time, and in 1989 Susan G. Cole created a perfect term—"pinkification"—in her book *Pornography and the Sex Crisis*. This term literally means "the act or process of being made pink or being saturated with pink." Cole's book is about sexual suppression and the subordination of women by men. Cole used "pinkification" as a reminder that the color pink and its implication of femininity and its spread in cultures was problematic and contributed to subordination. Cole was probably very familiar with Barbie dolls and their pinkification. Pink is a dominant color in Barbie's wardrobe, while her male counterpart, Ken, rarely has pink in his. Even astronaut Barbie (released in 1985) gets in on the act, with a white space uniform and pink helmet, pink moon boots, and pink bands marking the uniform around her waist, thighs, and ankles. Her oxygen tank is even pink. In a more recent example, JeongMee Yoon, a Korean artist and photographer, created the 21st-century *Pinkie* and *Blue Boy* with her art installation *The Pink and Blue Project*. She placed young children in their bedrooms with typical kid items, but for the male children the items were mostly blue and for females the items were mostly pink. The photographs are stunning and demonstrate Yoon's initial observation and stimulus for the project, which was that the phenomenon is tied to marketing, which specifically targets girls with pink and boys with blue. Currently the Pussyhat Project is popular. This movement recently found prominence in several Women's Marches across the globe. It was started as a symbol of unification; women at these marches wear pink wool caps topped with little pussycat ears. It is quite impressive to see these marches, where a significant proportion of women in the march don these hats to produce a sea of pink. As we write this book, the Pussyhat Project is under fire for using the color pink for a wide variety of reasons, including, but not limited to, the assertion that while pink represents femininity, it does not represent all women. Oh, and in 2019, pink is now the "new black," as it is used widely in suits, shirts, and ties for men. This fallout demonstrates further the fragility of color involved in CSVs.

Remember that we are discussing mostly Western cultures here. There are many parallel stories around the globe. For example, as marketing gets more and more global, the preference for pink as feminine and blue as masculine tugs back and forth at each other. We focus mostly on red (and its diluted cousin, pink) in this chapter to demonstrate the lability of color preferences in the history of some cultures. How do people use any of the other primary colors that humans perceive? A *cultural* history of color, instead of a *natural* history of one, would certainly make for a tremendous follow-up to this book.

Creating Colors

Color has been a key creative component in the arts. The ability to make colorful objects has been with humans for millenia, motivating generations of artists. Color can be described by three major properties. *Hue* is the color's name, such as red, blue, or green. (This aspect is pliable, as we have indicated earlier in this chapter, because names change with language and hence with culture.) The color's intensity is called its "colorfulness," or *strength*. The last characteristic of color in art is its value or *vividness*, meaning its lightness or darkness. Artists can alter the vividness of colors with several tricks involving lighting or perceived lighting. They alter hue with different sources of color. We have already mentioned ochre and cochineal as sources of hues, but the history of color usage in art is as expansive as works of art themselves are.

Art is as much a story of developing human consciousness, artistic capacity, and expression as it is a story about the history of color production. According to Victoria Finlay, author of *The Brilliant History of Color in Art*, over the ages different colors have been invented at specific periods of time by advancing cultures. The invention of colors at these specific times correlate with the beginning of a new artistic movement or some breakthrough in artistic ability. Her book is indeed brilliant, both in its illustrations and more importantly in the exposition of the development of art using color as a guide.

We have already mentioned the ancient *H. sapiens* use of ochre as a source of dark red for cave painting, but it can also be the source of yellow. Manganese, when treated properly, creates a deep black that was used extensively by cave artists tens of thousands of years ago. Chalk was also used, and this produces white. The ancient palette for the very first art produced on our planet was made up of white, dull red, yellow, black, and brown because the artists at that time were limited to using colors from nature. Most of the colors were from rocks and sand, where iron is deposited and where the other natural materials used by the cave artists were found. You might think that the art produced with such a basic and admittedly drab color palette would be rather dull. But those humans lucky enough to have crawled into Lascaux Cave in southern France or the caves mentioned earlier in this chapter are always stunned by the absolute beauty and motion these drawings hold. As more modern artists who draw or paint in monotones or in black and white prove, a limited color palette is not a drawback.

Red ochre appears to have been discovered over and over again by humans, more than likely because of its broad availability across the planet, so the palette for all parts of the globe—including Australia and the Americas—was pretty much the same. But different cultures used the palette very differently, as a comparison of European cave art and Australian Aboriginal art will attest. Different artistic styles are highly possible and do indeed reflect the cultural leanings of various populations. What is remarkable is that at this early stage in human history, with very limited contact among isolated populations, humans were engaging in art and expression of their cultures. The urge to make art, to discover colors, must have been overwhelming to these ancient *H. sapiens* populations.

As one might surmise, the range of colors missing in the palette were huge; no blue, no green, no in-betweens or shades of colors. It is not surprising, then, that these other colors were invented on the heels of cave art. Egyptian blue is an example of how persistent humans were at finding and making colors thousands of years ago. The paint made with this color is produced with great difficulty, and its production has many steps and precise temperature requirements. It was used by the ancient Egyptians in many of the murals in the pharaohs' tombs. It is a distinctive color that was lost for a long period of time, between the height of the Roman Empire and the color's rediscovery in the 19th century, because the recipe was lost, and it was so complex to make that it couldn't be copied. A similar color in a dye, Tyrian purple, was used almost ubiquitously by the Roman emperors. It was so desired by the emperors of Rome that they took it all for themselves by decreeing that it could only be used to dye togas of the Caesars. If you weren't a Caesar and wore this color, it meant execution. While this restriction was lifted a few hundred years later, wearing purple was taxed by the Romans. Ancient Romans expanded the color palette substantially. Up until the Roman Empire, ochre was the only widespread source of red, and it was a drab red, in the West. Two reddish hues—Cinnabar and vermillion—were probably invented in Ancient Rome. However, vermillion red is probably older (four thousand years), as the Chinese used it in some of their art.

Adding green, gold, ultramarine, and a wide array of other colors to the palette was a medieval and Renaissance innovation. Ultramarine is one of those colors, like Tyrian purple, that was difficult to make and hence fairly rare. It is made from a stone called "lapis lazuli," which is pounded and ground to a powder. Resins and gums from plants are added, followed by a complex series of kneading and filtering of the mixture. Ashes are added, as well as lye, and the concoction is placed in the sun to evaporate. Remember also that the bright reddish dye made from cochineal from the New World was recognized by the Spanish during the latter part of this period as an important dye. Painted art flourished during this period, and vividness (one of the big three aspects of color in art) became a major component of medieval and Renaissance art. It is also apparent that Arabian art included some of these colors, especially ultramarine and some green hues. The Renaissance masters expanded their color palette by mixing. These artists were the first sophisticated mixologists and created a broad array of additional colors with this mixing technique. While the palette appears to have gotten larger and larger, we still aren't near the full picture.

Post-Renaissance years saw the discovery or creation of a large number of novel hues for art. Finlay mentions longwood black, cobalt, lead white, indigo, Gainsborough blue (remember Gainsborough's *Blue Boy*), and rose as additions to the palette in the 1600s and 1700s. As the palette grew larger, the range of artistic approaches, subjects, and kinds of canvases also expanded. Still lifes and painted porcelain objects abounded. But equally important, commerce and economics of whole regions developed around some of these colors, much like the special relationship the Incans had with cochineal. Indigo is one of those color products. Originating in India, indigo is actually a very old color product, as it existed five thousand years ago in the Indus Valley. From there the product moved westward to the Fertile Crescent in what is now Iraq. The East India Company founded by the British traded in indigo. The French started a whole industry around this product in their colonies in the Caribbean. They eventually cornered the indigo market, essentially eliminating the British because the French product was superior.

During the 1800s, several novel hues were added to the palette, including titian red, Indian yellow, madder red, mummy brown, and graphite black (because this material was used for drawing). The most recent period of time added yet another onslaught of color, with mauve, Prussian blue, emerald green, sepia, manganese violet, chrome yellow, patent blue, tartrazine, rose Bengal, cadmium yellow, and lithol red. As you might have guessed, the names of these colors got more and more specific as they were added to the palette. In the 1920s the Colour Index International was adopted as a way to keep the names of colors consistent, first in the English language and now internationally. The index allows people to avoid the confusion that was caused by the immense number of names for colors being used and the inconsistency in how the names of colors had been used. The Colour Index™ was initially published in print but is now available solely in digital format (Colour Index™ Online at https://colour-index.com/about). First published in 1924, it is currently managed by two organizations with extreme interest in the consistency of colors—the Society of Dyers and Colourists (SDC) and the American Association of Textile Chemists and Colorists (AATCC). The index classifies color products using generic names. Currently there are twenty-seven thousand individual products entered into the database and about thirteen thousand Colour Index™ Generic Names. Both dyes and pigments are included in the database, and each product placed into the index is given a five-digit identifier. The systematization of color by the Colour Index™ might not be that important for an artist who is simply creating something beautiful. The colors that artists choose don't really need names in order for the artist to use them effectively. However, some artists might want to get in the ballpark, and in this sense the names generated by the Colour Index™ become important. The wide range of colors available to 20th-century and current artists allowed for a plethora of schools and styles to be created. Art was probably relatively uniform during certain ages because the color palette was limited.

The purpose of going through the evolution of color in art was not to discuss the aesthetics of art but rather to show how the colors developed from the first cave art to the modern era and how these opened the way for humans to expand the styles and meaning of art. It is clear that with the addition of a substantial number of colors to the artist's palette came novel forms of art created by the humans of that time period.

Parallel to using colorful objects and chemicals that reflect light, artists discovered the amazing properties of objects that glow or filter light. The stained glass windows in European churches are a most impressive sight. Generations of worshipers, whether they be Muslim, Christian, Hindu, or Buddhist, have marveled at the spectacular images that can be created and the impact they have. Small windows and lanterns were created by Egyptians and Romans. Syria and Damascus were centers of glass manufacture, and excellent work was done in the Assyrian Empire. The use of stained glass continues over centuries in Persia, and stained glass in Arab cultures is a highlight in the Islamic world. This is a global form of art that includes India, China, and many Asian cultures.

Making stained glass was a great breakthrough. Producing color by adding metal oxides or painting on glass provided a whole new dimension to artistic expression. The effect of glowing colors in a darkened room was and is something to be experienced. This technique lends itself to telling stories, to lifting the spirit, to displaying impressive ornamental geometrical decorations. Such uses of color have attracted humans for centuries; stained glass is still a major aesthetic attraction, even now and is still used in architecture, to produce an image or change the mood of the room. The trick is to absorb many of the colors from the spectrum of sunlight, or a candle or electric light for small objects. The technique of creating colors with stained glass is somewhat opposite to the creation of colors in paints. However, the range of colors for both techniques can be similar.

Over the history of art, new ways to use light and colors were invented and applied to create stunning effects on our collective human psyche. Besides the addition of new colors to the artist's palette, new technology has added to how humans make art. During the last 150 years, two really interesting and very precisely located shifts in color usage have impacted the arts. These are the addition of color to photography and cinema.

There is some controversy about when the first color photograph was taken. Perhaps one of the more famous candidates was made in 1861 by Thomas Sutton based on an idea from the famous scientist James Clerk Maxwell. It used a very crude RGB method, as it was a set of three photographs taken through red, green, and blue filters. These photographs were then projected through a colored filter corresponding to the color through which the photograph was taken. The three projections were overlapped precisely, producing the first color image made through photography. Up to that time, photography was displayed in black and white, or sometimes in sepia tones. This early reliance on simple black-and-white tones was a matter of technology. Color photography technology made some significant developments from the 1800s to the 1960s, when it really took off as a way to take photographs. There are literally hundreds of ways that color photography was implemented. It turns out that color photography was pretty much shunned for some time after its invention, more than likely because it was unreliable and expensive, and black-and-white photography remained aesthetically attractive to people who used it and viewed it. We quote the words of the mid-20th-century photographer who worked mostly in grey scale, Henri Cartier-Bresson: "Color is bullshit."

Color in cinema has been around a long time. Almost from the beginning of cinema, color was incorporated, sometimes by hand-painting the film, other times using early color film technology. There are almost two hundred full-length color films from 1902 (*A Trip to the Moon*) to the mid-1930s, including many silent films. Perhaps the most important development in color photography and cinema technology occurred in 1935, when the Kodachrome film process was invented. This subtractive process used three emulsion layers (cyan, magenta, and yellow) on film, which, when developed, gave a full color image. With this technological development, color made its way into some photographers' repertoire and into a lot of cinema. With respect to cinema, it is very unusual for black and white to be used today for a major motion picture, and television is almost all color except for those rerun stations. Some memorable departures from this preference for color cinema are Martin Scorsese's *Raging Bull*, Steven Spielberg's *Schindler's List*, the Coen brothers' *Man Who Wasn't There,* and the absolutely beautiful Spanish movie *Blancanieves* from Pablo Berger. A recent memorable black-and-white TV moment occurred when *The Walking Dead*, a gory TV series about a postapocalyptic zombie world, aired an episode entitled "Guts" in black and white. In many ways, cinema has the capacity to purposely use black and white and color to move audiences, and this in and of itself is an interesting addition to the palette of the moviemaker. Photographers have likewise taken advantage of the choice of a color palette versus a black and white one. One of the more visible examples of this approach in modern photography can be found in Robert Mapplethorpe's work. If there is anything to learn from the preceding cursory treatment of color in art and photography, it is that there will be more innovations, more colors, and more choices for the kind of media artists will use in the future.

In the last half century or more we have moved on—think about creating the effect of a stained glass window but changing the color in all places of a window at will to make a colored image. Being able to do this means that two new steps need to be created. The first is to divide the image into a huge number of very small parts. These small parts are now called "pixels." Now think about this: If we have these pixels, maybe we can use a control system to tell the pixel what to be. If we are in a black-and-white world it's easy. We just need to tell the pixel how much light to let through. Perhaps a number can be used to do this. A ten means block all light in the pixel and a zero means let the background light all come through. It turns out that numbers are a great way to do this for color images using pixels. For color, we divide the image into similar small parts or pixels and store the image as a sequence of numbers. In the case of color, though, each number describes the hue and the intensity of each pixel.

The second trick is to create a screen which can be controlled pixel by pixel. There are many versions: some have a white light behind and work like electrically controlled stained glass windows. We have them now as big flat screen televisions or in some smartphones and other devices with liquid crystal displays (LCD), with tiny pieces of liquid crystal for each pixel and white light behind. Alternatively you can make each pixel glow, with light emtitting diodes, plasma sources, or organic crystals. Every few years humans create a new technology, more colorful, brighter, and cheaper to manufacture the pixel effect. Welcome to the digital age, the 21st century, where sets of numbers are all the information required for amazing color images. Anything can be created, stored, and downloaded as numbers, so anything can be intricately colored with the very same numbers.

We now have displays and screens everywhere, and they produce colorful images. The many little pixels, which cannot be resolved by the human eye, produce the beautiful images we see on television screens and our phones. While they produce a huge range of colors, hues, and intensities, this actually is just a trick. These displays work basically with three colors only: red, green, and blue, abbreviated to RGB. These colors are designed to match our vision. For trichromatic humans, RGB is matched to the range of colors we can detect with the opsins our lineage has evolved. Just as we can see a huge range of colors through a combination of just three signals, we can produce the impression of a similar large range of colors by controlling the intensity of tiny pixels in red, green, and blue.

So, any cadmium yellow color you see on your computer screen is not really cadmium yellow, whereas the cadmium yellow that covers nearly half of the canvas of Wassily Kandinsky's *Impression III* is truly cadmium yellow. The pure yellow on your computer screen really isn't the same yellow as you would see in a painting in a museum. Wait a minute, isn't yellow one of those primary colors? It is for a rainbow that you see with your eyes, but computer screens use RGB technology. In each pixel there are three dots or bars that shine at different intensities at the same time, using the colors that most efficiently interact with the three different cone cells in our eyes: red, green, and blue. Your computer screen is made of millions of these pixels, each capable of shining the three colors at different intensities. Hence the statement that a computer screen only emits three colors is valid. The illusion of cadmium yellow or any color other than red, green, and blue is created by the immense number of pixels in the screen.

White on your screen is obtained by having the three dots or bars in a pixel shine red, green, and blue light at the same intensity. Black on your screen is caused by pixels without any of the bars or dots shining. Gray tones on your screen are created by pixels with the three bars or dots shining red, green, and blue but at an intensity less than full. Yellow is caused by pixels with the blue bar or dot turned off, and the red and green pixels on at full intensity. Cadmium yellow on your computer screen is a little like the normal yellow, only the intensities of red and green differ and the blue bar in the pixel is on in slight intensity.

When you look at a color picture printed out from a file on your computer, you are once again being tricked by what you see, only in a different way. Instead of RGB, it uses a different system called "CMYK," where the colors are cyan, magenta, yellow, and black, respectively. These are colors which can be combined to create any hue the artist likes. Again, the pixels in the printed picture are so small that we cannot resolve them with our eyes, so the actual colors reflected off the picture merge and create any hue.

The difference is that the CMYK color system for printing is what is called "subtractive," while the RGB system is additive. Since color printing from computers is usually done on white paper, the ink is placed onto this white background. Any ink placed on the white paper absorbs light at the wavelength of the color that the ink is. In this way the inks (CMY) placed on the paper are cyan, magenta, and yellow, because when red is subtracted from white light you get cyan, when green is subtracted from white light you get magenta, and when you subtract blue from white light, you get yellow. Ultimately, when all three colors (CMY) are present in equal and full intensity, you get black.

Additive RGB is so-called because the background is originally black (no light) and the three colors of the RGB scheme are added together to get the various colors that your computer screen shows. Ultimately when all three colors are present in equal and full intensity you get white. The full illusion in both of these processes is to create a wide array of colors that we think we are seeing but that we really aren't. In essence, if you print out that Kandinsky piece we mentioned above, you aren't really seeing cadmium yellow in the printout but mixtures of cyan, magenta, yellow, and black.

We are currently in a very colorful world, which can change in a millisecond flick of a computer command. Our world sends us images, information that then induces emotions and moods at will. When you look around yourself, can you distinguish between screens and the real world? Sometimes we can't. But do we have to? Color is central in this new digital world, for artists, for journalists, for everybody.

7

The Color
of Humans

No treatment of color would be complete without a discussion of color variation of humans. The colors of our hair, skin, and eyes all have interesting stories around them. Skin, hair, and eyes are the most obvious parts of our bodies with color, because these are all on the visible, outside part of us. Our skin is the largest organ in our bodies, and the range of color of this organ is spectacular. It is probably the single most variable organ on or in our bodies with respect to color. Eye and hair color also vary widely. But color abounds in other parts of our bodies, specifically on the inside. Our blood is both blue (or is it?) and red, our internal organs vary in color depending on how well they are working, and our bones and muscles have distinctive colors too.

Blood and Guts

Blood moves oxygen around our bodies. The hemoglobin in our blood cells does this job. It is made up of four subunits that bind to oxygen and uses an iron atom for the binding, while the heart pumps blood through the body. When iron interacts with oxygen it produces a rusty red color; the more oxygen, the deeper red the color that is emitted. As the hemoglobin is moved around the body via our circulatory system, the oxygen is dropped off in the various tissues that need it to function. The hemoglobin then circulates back to the lungs, where it picks up more oxygen, goes to the heart, and repeats the circuit. Not all hemoglobin molecules drop off oxygen in tissues, so some molecules on their way back to the heart and lungs will still give off a red color. Others without oxygen lose the red color. The combination of the hemoglobin with and without oxygen produces a dark deep reddish-purple color that appears blue to our eyes through the skin.

Our inner tissues have different hues depending on what they do and how they are doing it. You wouldn't guess it, but your liver can take on a rainbow of colors depending on its condition. The liver filters blood that circulates around the stomach, which has picked up nutrients and other things, including alcohol if you have been drinking. Typically, a healthy liver is its own color called, not surprisingly, "liver." This color name was first used in the English language in the 1600s and is between red and brown. One of the main conditions that will change the color of the liver is jaundice. There are many ways to become jaundiced, all of which are manifested by a yellowish discoloring of the skin and whites of the eyes, to name two outer tissues impacted. The yellowness comes from a molecule made by the liver, bilirubin, a pigment that gives off the yellowish color. The liver makes bile, a fluidic side product produced by breaking down senescent red blood cells; this bile contains the bilirubin. When the liver overproduces bilirubin, the yellow coloration occurs in the skin and whites of the eyes, and the liver takes on a yellowish or greenish color. A liver that has overworked itself because of alcohol will appear brown, and even a very dark brown if the blood has been poisoned with too much alcohol. Emphysema, a condition of the lungs caused by malfunctioning air sacs, will cause the liver to turn black, and cancer of the liver oftentimes produces white or gray sections of the liver.

Because the lungs oxygenate blood as the hemoglobin containing molecules pass through them, the lungs are a bright pink at birth, but they get grayer and darker with the aging process as more and more junk from the air is breathed in. A healthy adult lung will be between a pinkish brown and gray, but smokers' lungs will be much darker, even black, with deep black patches where tar has accumulated. Cancerous lungs take on a blackish hue. We already have mentioned emphysema and its impact on the liver, but this condition also turns the lungs black due to the accumulation of red blood cells. Kidneys are deep red in healthy humans because the job of the kidneys is to filter stuff out of the blood to remove waste and excess water. The waste and water are sent to the bladder in urine, so what comes out of the kidneys is actually more interesting with respect to color. Your urine should be a light yellow, but if it is dark yellow it probably means that you are dehydrated. Sometimes your urine can turn blue, and this is caused by the passing of blue dyes used in food products through the digestive system and kidneys. This could indicate a malfunction of the kidneys, which should be filtering out the blue dye but aren't. Urine can also have a pinkish or even a red color, which could indicate two things. Reddish urine might simply be caused by the passage of red dyes, such as beetroot, by the kidney into the bladder. But reddish urine can also indicate blood in the urine, which is not good and probably indicates some form of infection in the bladder, urinary tract, kidneys themselves, or a disease of the kidneys. Our pancreas has a dual function—secreting endocrines that regulate blood sugar levels and aiding digestion. It is a tannish pink in healthy humans and influences the color of the urine and feces too. The urine can turn a deep brown with a malfunctioning pancreas, and the feces will seem a chalky gray.

The gall bladder is a tiny organ that many of us can do without. It stores bile produced by the liver, which is used in digestion of food, especially fat. Typically, the gall bladder is empty after a meal but swells to the size of a large olive before a meal because it is full of bile. The gall bladder is dark green because of the bilirubin pigment it stores. It will turn various colors if it is diseased or malfunctioning. Both the kidney and gall bladder when not functioning properly will produce deposits commonly known as *stones*. Since gall stones are of two types—cholesterol and pigment—they are of two colors. Cholesterol gall stones are caused by the accumulation of cholesterol and are yellow in color. Pigment gall stones occur when the gall bladder accumulates too much bile or bilirubin, and this precipitates as a brown or black solid. Kidney stones are amazingly colorful but intensely painful because they are, after all, crystalline deposits in the kidney that get passed on to the bladder and then excreted. Their broad range of colors are the result of there being three major kinds of stones. Calcium oxalate stones, uric acid stones, and struvite stones give off the different colors. The combination of calcium and oxalate together produce a dark brown or gray colored stone, while uric acid stones are brownish-white. Struvite stones are the most colorful in that they are pinkish and yellowish.

The intestines are rather nondescript with respect to color, as they are a basic light pink. Your rectum is the same pinkish color, but, like urine, it is the poop that takes on a rainbow of colors. Feces are made up of stuff our digestive system did not process, like cellulose from plants and tons of bacteria, along with a little bit of yourself, because as you poop you slough off cells from various parts of your digestive tract. In fact, poop is used extensively in forensics, as it can be used to identify individuals. As any parent can attest, infant poop is both colorful and disconcerting. An infant will excrete everything from yellow to red to green to brown and even blue, all at different stages of infancy. Usually this rainbow of colored poop is nothing to worry about and is caused by the infant's digestive system coping with exposure to a plethora of bacterial species that pass through their bodies.

Though poop is often written off as just being "brown," color is actually an important aspect of poop, because poop color can mean so many different things. Yes, the normal, average color of poop from a healthy human being is indeed brown, yet adult humans will excrete a rainbow of colors of poop, the majority of which are quite normal. As with babies, most of the off-color adult poop is not a sign of disease or a disorder. Blue adult poop probably comes from something you ate that had blue pigments in it, like blueberries. While red poop can mean bleeding, it could also mean that the red pigment of that beet salad you had the night before just made it through your digestive tract. But there are some colors of poop that should be considered troubling. Black poop could be the result of eating a lot of black licorice, or it could be the result of blood in your stool caused by some internal bleeding in the stomach or small intestine. We have already mentioned beets as a cause of red poop, but this color in the feces could also mean that there is bleeding in the lower digestive tract, the rectum, or the anus. Green poop probably means you have been ingesting a lot of plant matter like kale or spinach. But it can also mean that your food is passing through your digestive tract unusually fast, which could be troublesome. Grayish or pale-white poop can mean that you are not excreting bilirubin, which in turn could mean that you don't have enough bile, perhaps connected to your gall bladder misfiring. All of these colored poops more than likely have a nonserious explanation, but if your poop is bright yellow and greasy this means that fat is being excreted at an unusual rate, which should be alarming. It could also mean that there is some kind of malabsorption disorder like celiac disease. Celiac disease is particularly illustrative of how a color gets its way into poop. It is an autoimmune disorder triggered by gluten, where this molecule causes an overreaction of your immune system. Your small intestine has small projections along its inner passage (called "villi") that increase the surface area of the intestine. They get damaged by this overreaction. When this happens, the small intestine malfunctions and the food passed through isn't properly digested, leading to the yellow color.

Even our bones have a color story to tell. While the general impression is that bones are white, they don't necessarily need to be. We have this notion perhaps because when we examine X-rays, they look white on the black background of the X-ray. Alternatively, most bones we see are in fixed skeletons, whose bones have been bleached white. Bones can actually be a light yellow, light pink, or light brown, but the color is most appropriately called "off white." The muscles of our body are pinkish in hue because they are fed by blood vessels. While a human brain when preserved can look yellow or dull yellow, it is far from that color in our skulls. On its surface it is a grayish color, which has tinctures of pink as a result of the blood supply to the brain. If one sections a newly removed brain though, the gray on the outside gives way to white and also to very dark patches that are even black. The gray color represents what is commonly referred to as the gray matter of our brain, where nerve cells congregate and control muscle movement and light and sound perception. The white matter of the brain is white because it is chockful of axons coated in myelin, which gives off a white color. Deeper into the brain are other gray areas but also very dark patches of black, such as the substantia nigra, which is black because of an extremely high concentration of neuromelanin in dopaminergic neurons.

The rainbow of colors inside us is impressive, but these colors are almost never used to characterize humans in practice. One never hears "Oh, did you see that guy with the yellowish-green gall bladder?" or "That woman with the very gray brain is listening!" Instead we use the colors of skin, eyes, and hair to do the majority of communicating about the appearance of people. This way of looking at people has been a hallmark of humanity but also has wrought some of the most dangerous, insidious, and damaging aspects of our history.

The Human Color Palette

Carolus Linnaeus (Carl von Linne), the great Swedish natural historian of the 1700s, characterized and named over ten thousand species in his

Systema Naturae. Our own species, *H. sapiens,* was named by him, and if he had quit there, perhaps we would have been better for it. But he went a step further and divided *sapiens* into four major divisions that roughly aligned with the racist European notions of people on our planet. He based these divisions on skin color. When taxonomists name things, they give them a binomial with the genus name first and the species name second. Taxonomists oftentimes use the Latin words for an organism's color or appearance in the binomial. For instance, *Quercus albus* is an oak familiar to many North Americans—a white oak, to be exact—where the genus name *Quercus* is followed by the species name *albus,* which is Latin for "white." Linnaeus named the four *sapiens* subdvisions *Europaeus albus, Americanus rubescens, Asiaticus fuscus,* and *Africanus niger,* corresponding to continental origins of these populations. Within their formal descriptions, Linnaeus used these skin color designations: Europeans were *albus,* or white; Native Americans were *rubescens,* or red; Asians were *fuscus,* or brown; and Africans were *niger,* or black. In furthering the racist descriptions, he formally described each of these with a behavioral aptitude. So, Europeans were "governed by laws," Asians were "opinionated," Native Americans were used to "customs to govern behavior," and Africans were "impulsive." Note that he named not only four different species for differently colored humans but four genera as well, and in the process the taxon *H. sapiens* was trashed. The next two hundred years saw a lot of taxonomic wrestling over how we should name our species, whether combining several into a single taxonomic name or splitting them into several. Today there are no valid subspecies of *H. sapiens* and only a single species name for all of the people on the planet. Given that species in a biological sense means an interbreeding group of organisms, this is a good descriptor for humans.

We can't really blame racism all on Linnaeus, as human scientists, politicians, and philosophers have had ample time to correct the problem. Skin color is integral to the mistaken view of biological human races, so we delve into this topic in some detail here. Ultimately, we probably should blame the sun for the race problem, because it is UV radiation from the sun that is at the core of skin color differences. Anyone who lives in the tropics can attest to the intensity of UV light and its dangers. Low-latitude environments such as Central Africa, South America, and parts of Asia are flooded with more UV radiation than northern latitudes. The average Global Solar UV Index (UVI) at the equator is more than twice the maximum UVI in northern latitudes. Twice as much UV radiation (UVR) might not seem like much, but it certainly has had an impact on how skin color has evolved in ours and other species. UVR radiation has many effects on our bodies, the most significant being skin disorders, including skin cancers. Most species solve the UVR problem with hairy bodies, but our species has solved it with melanin, that protein we discussed in the context of mouse coat color in chapter 3. In fact, certain levels of melanin in the skin can absorb more than 99 percent of the UVR hitting a body.

You might be asking, Why did we lose our hair? Certainly the ancestral state of our lineage was hairiness. We don't know exactly when our lineage lost the hairy body phenotype, but two million years ago is a good estimation. This is about the time ancestors in our genus moved out of the forest onto the savanna. At that time an evolutionary swap meet occurred. By moving out of the forest, humans left themselves vulnerable to overheating if they had a heavy coat of hair. This is probably why we don't see chimps and gorillas romping across the savanna. For humans, the development and safety of the brain was tantamount, so overheating in the shadeless environment was an important environmental challenge to the early ancestral human populations. If the brain overheats even slightly or for a short period of time, it can have severe and lasting effects. Losing hair was a way to cool the body and keep the brain from overheating. Our ancestors evolved a way of secreting sweat through cutaneous sweat glands. The sweat would then evaporate in the air and hence do the cooling. Here is the swap meet: losing hair to allow for the sweat to do its job resulted in our skin being exposed to the sun. But because our ancestors' genomic makeup included control of melanin, the trade-off resulted in increases of melanin in the skin of these ancestors and their descendants—us. Our ancestors in Africa were very dark skinned then to reduce the impact of UVR.

But today we see a broad range of skin colors on the planet, and this begs an explanation too. The relationship of skin and UVR is actually more complex than the story we presented above. UVR is both a stimulatory and inhibitory interactor with our bodies, in that it regulates many biochemical reactions in the body that maintain physiological stasis. Long wavelength UVR inhibits folate, an essential vitamin that is involved in DNA replication of cells. UVR has a huge effect on DNA and its replication in other ways too. UVR at the short wavelength end enhances the synthesis in the skin of vitamin D, which is needed for proper calcium metabolism in the body. There is a teeter-totter, so to speak, between the enhancing and inhibitory effects of UVR. Hence a balance between the two effects of UVR evolved and continues to evolve. It has produced two global gradients, or clines, of human skin color. One cline involves a response to the inhibitory effect of UVR and runs from the polar regions to the equatorial regions. This gradient would favor the inhibition of folate. Remember that folate helps in DNA repair caused by UVR. The other gradient runs from the equator to the poles (the opposite of the first gradient) and favors the enhancement of melanin production. Both of these gradients favor darker skin around the equator and lighter skin toward the poles. As humans migrated out of Africa and toward the poles, dark-skinned ancestral populations evolved into lighter-skinned descendant populations, because enhanced melanin production, and hence darker skin, was not selected for in those populations.

Remember though our discussion of Lewontin and Gould's spandrels. We don't really want to fall into the adaptationist trap that we described in chapter 5. While it is pretty obvious that increased melanin and heavy pigmentation are indeed adaptive in the tropics, why lighter skin is predominant toward the poles is not at all as clear as the gradient story told above. There are a couple of reasons why lighter skin might have taken hold in nontropical regions. One hypothesis is that maintaining higher levels of melanin in the skin comes at too high a physiological cost to the bodies of humans living at higher latitudes. Another hypothesis would be that there is selection on Vitamin D synthesis in the bodies of humans living at higher latitudes, which would then dovetail with lighter skin phenotypes. Or finally, we might be seeing simple relaxed natural selection for dark skin in higher latitudes. Whatever hypothesis best explains the trend, it is not under debate that the pattern is there.

The genetic basis of skin, hair, and eye color has been the subject of research for over a century. The three major ways that researchers approach the problem are through study of Mendelian recessive mutants that affect skin color, through phenotypic studies of skin color, and through the recent novel approach GWAS. The Mendelian studies are generally done on albinism, which is usually a recessive Mendelian inherited trait. Albino individuals completely or partially lack pigment in their skin, hair, and eyes. In addition, people who have this syndrome have several vision problems as a result of the enzyme deficiency caused by the recessive allele.

The genetics of albinism is a classical story in genetic discovery. While Gregor Mendel discovered the basic principles of genetic inheritance, the development of the discipline led to new tools for studying variant phenotypes. Thomas Hunt Morgan, who won the Nobel Prize in 1933 for his work on how chromosomes are involved in heredity, established this rule using *Drosophila*, a small fly. He used easy-to-interpret phenotypes that were inherited in a Mendelian fashion. This means that he used phenotypes, or traits, of his flies that were caused by changes in single genes that could be interpreted in the same way that Mendel interpreted pea heredity. Indeed Morgan "found" his first variant, a white-eyed fly (*Drosophila* have red eyes in nature) in lab strains he was working with. To find more and to accomplish his seminal work in genetics, he established a lab now affectionately known as the Fly Room at Columbia University, where he was a professor. His Fly Room is where some of the most famous geneticists of the 20th century cut their chops. Several started out as bottle washers (the flies were reared in small milk bottles and were reused to save money, so they needed to be washed occasionally). Hermann Joseph Muller, another Nobel Prize winner for genetics, was one of several Columbia undergrads who worked in the Fly Room, which is now a men's bathroom on the fifth floor of Schermerhorn Hall on the Columbia campus. After earning his PhD, Muller went on to work in Texas, where he studied the impact of X-rays on *Drosophila* phenotypes and was able to rapidly generate *Drosophila* mutant phenotypes. The number one rule for genetics established by Morgan and Muller was to either find or generate a variant. Once the variant was established in culture, they could then manipulate it in crosses and discover the genetic and physiological basis of the trait. Once the inheritance of the trait was figured out, biochemistry and physiological experiments could be conducted to figure out the full phenotypic spectrum of the changed phenotype.

Albinism has been critical in pinpointing the various genes involved in skin color because it is one of those "found" mutations. Of course, it would be unethical to generate mutations in humans, so all of the early work on Mendelian traits in humans was accomplished after finding variants like albinism. It turns out, though, that albinism in humans isn't a single Mendelian trait but several. In fact there are seven genes that, when altered at the DNA sequence level, can result in Mendelian inherited albinism. They are named OCA1-7 (OCA stands for "Oculo-Cutaneous Albinism"). The genes involved in these seven different Mendelian inherited albinisms include genes that code for proteins directly involved in the synthesis of the two major kinds of melanin found in skin cells that are involved in pigmentation of skin. We have explored the physiology of melanin production in the context of coat color in mice in chapter 4. Remember that the two major products of the pigment cells in mouse coat color are eumelanin and pheomelanin. In humans there are actually two kinds of eumelanin, based on the biochemical pathway they get produced by—DHI eumelanin and DHICA eumelanin. All three of the human melanins start out as two small molecules: tyrosine and dopa. These smaller molecules are combined together through several chemical reactions in the keratinocyte, a specific kind of cell near the outer layer of the skin. The first few steps in the synthesis are critical, and mutations in these steps cause albinism. Several genes that produce enzymes are involved in the synthesis of eumelanins and pheomelanin in the keratinocyte, three of which are tyrosinase (*TYR*), a tyrosinase-related protein (*TYRP1*), and dopachrome tautomerase (*DCT*). Mutations in any one of these genes will result in albinism. For these kinds of albinism, the genetics and the physiology are well understood. In people without albinism, these three kinds of melanin are properly made in small cellular bodies called "melanosomes." The precursors of eumelanin and pheomelanin in our skin cells need to be transported to the melanosomes for the production of skin pigments and back to the keratinocytes, where the melanin ends and where the skin color phenotype is ultimately expressed.

By understanding the tanning process in humans, much has been learned about how skin color is controlled at the genetic and molecular level. A veritable alphabet soup of genes is involved. Here is a list of some of them: *CPD*, *POMC*, *KITLG*, *MC1R*, *MSH*, *PKA*, *CREB*, *MITF*, *MAPK*, *HBD3*, *ASIP*, *IRF4*, and *TYR*. In the keratinocyte and also the melanocyte, these genes become activated when exposed to the sun's UV damaging radiation. Suffice it to say that these genes produce proteins that assist in transporting precursors to the skin pigments that ultimately reside in the keratinocyte as a shield around the nucleus of those kinds of cells. If any of these genes are knocked out by mutations or their expression is altered by mutations, then melanin does not get properly produced or transferred to the keratinocyte. We already know a lot about a few of these genes from studies on albinism. *TYR* and two other of these genes—*ASIP* and *MC1R*—should be familiar to us in that they are involved in the coat color melanism we explained in chapter 4.

Other genes involved in skin color have been discovered through genomics, where the genetic information from thousands of genes is assessed for whether the genes are involved in skin color. As we mentioned above, there are two genomics approaches that are used: the first being phenotypic studies followed by genetic mapping, the second being GWAS. One of these approaches is a very statistical one; GWAS studies use large numbers of individuals and strong phenotypic data to determine which genetic variations in which genes are involved in skin color. The end product of these kinds of studies is a statistical statement about how much of the trait's variance can be explained by the identified genes. These approaches—genetics of albinism, phenotypic studies, and GWAS—together have uncovered an even larger number of genes involved in skin pigmentation, enlarging the alphabet soup of genes involved. There are over twenty implicated genes and hundreds of different DNA sequence variants (positions in the genome with different bases than a reference genome) involved in skin pigmentation. While this is by no means a remarkably large number of genes, it still makes skin color a very genetically complex phenotype and not at all as simple as the melanism examples we discussed in chapter 4. William J. Pavan and Richard A. Sturm in their recent scholarly review entitled "The genetics of human skin and hair pigmentation" suspect that "more genes will be identified through the study of pigment variation in Africa."

To demonstrate how complex the control of skin pigmentation is in humans, we look here at studies of skin color in Africans. In a study of over two thousand Africans from several countries, two genes previously identified from albinism studies (*SLC24A5* and *OCA2/HERC2*) were confirmed by GWAS, and another two (*MFSD12* and *TMEM138*) were newly found. These genes are particularly important in explaining nearly 30 percent of the genetic variance in skin color, with *SLC24A5* explaining the majority (13 percent) of the variance, and hence being tagged as a major locus in skin color determination. One variant in this gene is found in very high frequency (28–50 percent) in the three African populations examined and shows interesting levels of variation in areas outside Africa.

The somewhat tight association of *SLC24A5* has been used to examine the flow of skin color genes throughout the globe. This gene produces an enzyme that aids in the transport of melanins across the membranes of cells. How efficiently the transport of molecules across membranes is carried out is a unique way that organisms have evolved to control the levels of molecules like melanin in cells. Because only a little over 12 percent of the variation of skin color is tied up in *SLC24A5*, the overall interpretation of the genetics of skin color is very difficult when just considering this single gene. However, a close look at the gene and its population dynamics is illustrative of some very basic aspects of how skin color is maintained in populations. Nearly every African in the populations examined has a certain form of that variant we mentioned above. This variant is in a coding region of the gene, so the variant is in a codon that codes for alanine. Most Europeans have a different variant in the DNA sequence position that changes the codon to code for a threonine. Simply put, the alanine and threonine in this position influence the proteins' activity in different ways and, as we point out, account for a considerable amount of the variation among these two groups of people. If the two continental groups of people were the only humans on the planet, then the story would be very simple and *SLC24A5* would explain skin color differences. But there are indeed other people on the planet from other regions. It turns out that 93 percent of very heavily pigmented Africans and much lighter-skinned Asians have alanine in this position. Clearly Europeans and Asians have evolved different genetic pathways to be light skinned. And keep in mind that there are also a significant number of light-skinned Africans as well as dark-skinned populations elsewhere on the globe.

When populations of people of mixed African and European ancestry are examined, the story gets even stickier. As with the initial study on *SLC24A5*, the researchers measured the amount of melanin in their subjects' skin while getting sequence information on the *SLC24A5* gene. The result of the study indicated that people who have homozygous genes for the alanine variant had more melanin in their skin than people who are homozygous for the threonine gene. This information might get us to fall into a "this is simple" mode. But most populations aren't made up of just homozygotes of the two variants present. Don't forget there are a large number of heterozygotes in such populations that arise as a result of Mendel's rules of inheritance. These heterozygotes tell us that only 38 percent of the variation in this mixed-ancestry population can be explained by the variants in *SLC24A5*. This is not too different from the original study on the three African populations, and it is not surprising that different levels of variance can be explained in the two different studies. These results mean that a large proportion of the variance (greater than 60 percent) must be attributed to other genes and gene complexes. It also means that there is more than one way to have gene combinations to produce low amounts of melanin and high amounts of melanin, but even more ways to have in-between amounts of melanin. These in-between ways are accomplished by the interaction of the many genes we have discussed so far.

Hair and eyes

Hair color has a similar story as skin but is even more complex. Recall that a hair follicle is chockful of keratinocytes in most of the follicle. The base of the follicle is comprised of dermal cells, which communicate with melanocytes that provide melanin to the developing hair shaft. There is a lot of signaling via proteins coded for by the genome that modulate pigment production in the follicle; because of these interactions, the determination of hair color is thought to be more complex than skin or eye color. Despite this complexity, hair color can be predicted more accurately because it has a high heritability (97 percent), making it one of the more heritable outer visible traits of our bodies. One GWAS study of hair genetic architecture revealed over 120 genetic loci involved in the trait. The study included data from more than 300,000 individuals listed with 23andMe and UK Biobank. Prior to this study only twenty-eight genes were thought to be involved in hair color, so many of the genes found in the GWAS study are new and could also be involved in skin color and eye color. Another equally large study implicates 163 distinct genes and 200 single nucleotide polymorphisms, or SNP, variants in hair color phenotypes. Each SNP variant represents a position in the genome where the nucleotide in the SNP position is different amongst humans. These SNPs are involved in the continuum of hair color from blond to black hair. While there is some predictive capacity for hair color, as we will see, eye color is more predictable, but it is safe to say that hair color is highly polygenic.

For eye color, there are ten to fifteen genes involved in producing pigment in eyes, and fewer than thirty or so DNA sequence variants involved. This makes eye color a little less difficult to understand than skin or hair color, but still very complicated, as eye color is not a simple Mendelian trait. Eye color varies as a function of the ratio of eumelanin to pheomelanin. For this reason, not surprisingly, eye color uses some of the very same genes that are involved in skin color. There are melanocytes in the back layer of the iris, called the "stroma." The big difference between the melanin molecules in the iris and those in the skin and hair, as we will see, is that in the eye, the pigment molecules reside in the melanocytes and are transported elsewhere. Since the spectrum of eye color runs from blue to green to amber to brown, it is instructive to look at the ends of the spectrum—blue and brown. Blue eyes have a greater than ten times amount of pheomelanin than eumelanin. Brown eyes have roughly equivalent amounts of the two melanins in the iris. As we mentioned above, albinism causes several ocular problems, but it also can produce irises with little or no pigment, so the blood vessels of the eye show through in a big way. This lack of pigment causes the eyes of someone with albinism to have red or pinkish eyes. People with very dark eyes essentially don't have black irises but rather deep brown eyes and so are a deep variant of brown.

As we said above, blue eyes are the result of the ratio of pheomelanin to eumelanin, and green eyes and hazel eyes are controlled the same way, but with different ratios of these two melanins. But it is not as simple as just the ratio of melanin. The eventual color also involves the optics of light in a suspension or colloid. If you fill a glass with milk and flour, a strange bluish hue will emanate from the glass. This is because of the phenomenon Tyndall scattering. The stroma is essentially colloid-like with respect to the melanins deposited there. Blue skies (and their capacity to change to red and yellow at sunset) and blue icebergs are the result of similar but not identical physical optical properties—Rayleigh scattering. Some older icebergs are blue because as they age, they eliminate air from their inner areas and become very different kinds of suspensions of ice than younger icebergs. When sunlight hits the older icebergs, longer wavelength light is absorbed and the blue and green wavelengths are reflected. If not for the Tyndall effect, our eyes would range from dark brown to very light brown. It is important to realize that there are no red or green or hazel pigments in the iris in people with these eye colors. Eye color is yet another example of structural color—that phenomenon we discussed when we examined organisms like the iridescent beetles in chapter 5. People with blue and green and hazel eyes actually exhibit changes in the blue, green, and hazel eye colors as a result of the structural color phenomena in the eye.

As we mentioned, the genetics of eye color is complex. While at one point in time blue eyes were thought to be a simple Mendelian recessive trait, it turns out that because of the large number of genes involved in the expression of eye color, the situation is much more complex. So complex that almost any eye color can arise from a mating. But can we predict what a person's eye color will be from their genes? Fan Liu, Manfred Kayser, and their colleagues suggest that it should be quite easy to do if a large enough number of genetic loci involved in eye color are used and a large number of variants for these genes are involved. They took eight genes (of the at most fifteen) involved in iris color, used thirty-seven variants (an average of four variants per gene), and looked at the variants for over six thousand Europeans. They used these thirty-seven SNPs, of which thirteen were removed because they showed linkage to variants in the rest of the sample. Of the twenty-four remaining variants, only the variants from six genes—*HERC2*, *OCA2*, *SLC24A4*, *SLC45A2*, *TYR*, and *IRF4*—are needed to make accurate predictions of eye color. They constructed what are called "receiver operating characteristic" (ROC) curves for each variant and different combinations of variants. Since each individual has a known eye color, the researchers can check to see how well their predictions match the actual eye color of an individual in the study (the true positive rate, or TPR). They can also see how well their predictions don't match (the false positive rate, or FPR). They then produce the ROC curve by plotting the TPR versus the FPR for different levels of significance. If there are only true matches, then the curve will be a straight line and the area under the curve will be 1.0, indicating a 100 percent agreement of prediction with reality. As the number of TPRs taper off, the FPRs go up because the model isn't predicting the eye color as well, and the area under the ROC curve (AUC) will decrease incrementally from 1.0. The area under the ROC curve is roughly correlated with the predictability of the model. The model for brown eyes using this approach has an AUC of 0.93, for blues eyes an AUC of 0.91, and for in-between irises, an AUC of 0.71. Adding more variants slightly increases the accuracy of the models for the six strongest variants. In fact, nine of the variants have little or no impact on the AUC, or, in other words, on the predictability of an individual having brown or blue eyes. What this means is that information on the six variants (SNPs) from the six different genes listed above is all one needs to determine whether a person will have brown eyes, blue eyes, or in-between eyes. While the predictions for in-between eye colors are lower, brown and blue eyes can be effectively eliminated as a potential color for such individuals. The success of Fan Liu, Manfred Kayser, and their colleagues is important for three reasons. First, it means that these six markers can be used in a forensic context. Second, it means that complex traits or phenotypes with similar genetic architectures (the same number of genes and degree of interaction of the genes) can be deciphered using the approaches they outline. Finally, connected to the forensic implications of the approach, scientists can predict the eye color of long-dead humans like Neanderthals, Denisovans, and prehistoric *H. sapiens* because the genomes of these individuals have been sequenced. Two Croatian Neanderthal females, for instance, were predicted to have brown eyes. Another ancient human girl, this time a Denisovan individual, was predicted to have brown eyes also. Contrary to the classical view of these non-*sapiens* humans having fair skin and hair and blue eyes, it is more than likely that they had brown eyes. Using these same individuals, geneticists predict that they also had brown hair and a range of skin color. This doesn't preclude the occasional blue-eyed/green-eyed individual with red hair as a result of mutation in the many genes involved in pigmentation in humans. Studies of ancient *H. sapiens* populations that sequence their genomes indicate strong selection for the *HERC2*, *SLC45A2*, and *TYR* genes in the melanin pathway, suggesting strong selection for eye color, hair color, and skin color in line with the clines we discussed earlier in this chapter. Genome-level information from populations of ancient Europeans indicates that the genetic changes necessary for blue eye color more than likely arose between fourteen thousand and seven thousand years ago. This means that most humans until about ten thousand years ago had brown eyes, and that lighter-colored eyes are very modern evolutionary inventions.

Resistance

While skin, eye, and hair color might ultimately be predictable from genome information, we humans have to resist the tendency that we have to accentuate the differences between people and not the similarities. Almost all racism starts with a misguided conception that there are significant differences between groups of people. While human skin color can be different from region to region, there are three things that render it irrelevant to perceived differences between the so-called races of people we discussed at the beginning of this chapter when we brought up Linnaeus. First, skin color is marginally, if at all, connected to other phenotypes (other than hair and eye color). While some of the genes involved in skin color might also be involved in other traits (and indeed they are to a big extent in eye and hair color), their involvement in other traits is so minimal as to be ignored. In other words, skin color has nothing to do with traits that racists associate with groups who have skin color different from their own: things like low intelligence, threatening demeanor, athletic ability, irresponsibility, sociality, crime, and the like. Second, skin color and its cousins eye and hair color have complex genetic architectures. There are many ways to make white skin and many ways to make black skin and these involve incredibly complex genetic interactions. We might be able to decipher the details eventually, but they are not significant and would simply split us into smaller and smaller groups, which wouldn't reflect reality, given our modern propensity to move around the globe.

Finally, skin, eye, and hair color are adaptive and hence have undergone rapid and fairly directed evolution enhancing those traits as visibly different. But other traits, such as the ones that racists harp on, are less prone to the directional and clinal selection that pigmentation has experienced, rendering racist arguments moot. On the other hand, the rapid process of cultural evolution might be more aligned with the rapid rise of skin, eye, and hair color adaptation. It is well known that while the biological context of race is fallacious, the cultural context is fully understandable and more significantly important. The importance lies in using culture to explore and explain our differences and not some underlying fallacious biological basis.

Instead we find it remarkable that the determination of these outwardly visible phenotypes has the similarities that they do. By simply tweaking the pigment genetic toolbox, nature produced the amazing degree of variability in outward human coloration that we see on our planet in our single species.

The variation in human populations and our capacity to use this variability to adapt to different environments is the real spectacle. The speed of these evolutionary transitions only points to the insignificance of other perceived differences among humans. In summary, we are simply a highly variable species.

8

The Color of Our Minds

Throughout this book we have tried to show how color has influenced not just human existence but the entire natural world around us. One of the reasons that color influences human culture and behavior so much is that our minds are full of color and color is full of our minds. It is hardwired into our very biology. This chapter examines the moodiness of color and how color has perfused our minds so much that perhaps it might be a key to explaining how we became the conscious beings that we are.

Color Moodiness

We have discussed how colors have been integrated into the many cultures on the planet and how these colors take on different meanings in different cultures. Red is one example of a color that has very pointed meanings across a wide range of cultures throughout human history. While there are several strong discrepancies between the way different cultures react to color, there are also some universals. For instance, it has been suggested that humans are aroused more in response to light of long wavelength (red) than light of shorter wavelength (green and blue). Part of the problem of determining the role of color in our emotional makeup is how to measure our emotions. Experiments that psychologists conduct to address color emotion have to be tightly controlled, or they simply don't tell the researcher anything about color and emotion. The simplest experiment is to ask a participant in the study to name the first color that comes to mind when a specific emotion is mentioned. The experiment can be reversed by asking the participant to name the emotion that first comes to mind when a specific color is mentioned. Oddly enough, these two ways of obtaining data, while seemingly aiming at the same thing, are not equivalent. For instance, a participant may associate red color with the word *angry* and the word *angry* with red color in the converse experiment, but because the possible answers to the first question (anger, sadness, happiness, etc.) differ from the possibilities for the second question (red, black, blue, green, etc.), the two questions do not address the same concept. Little things like this example harass the researchers when doing this kind of work, making it a difficult endeavor and one that requires a huge amount of forethought and control.

Color emotion correlation studies have been carried out now for eighty years, but with no major breakthroughs other than to describe the phenomenon. Its connections to neurobiology and color physiology are still to be made. But the study of color emotion correlation has become very important in the field of neuroeconomics, as we will shortly see.

The design of such psychological studies is by no means easy. It requires great attention to detail, and the studies can lead to inconclusive outcomes. For example, many studies that examine color preferences and correlations with emotions are about students at a university, probably because students offer a cheap and efficient resource for data collection (through an offer of beer money or extra credit in a course). In one study from the Philippines, the tables were turned and faculty members became the subjects of a study by their students. The study aimed to understand personality and gender differences in the context of color preference.

The results look interesting: the students found that there was indeed a gender difference in color preference. Most males preferred blue, while most females preferred red. While there were some departures from this general result, this simple survey of about 100 faculty members reinforced the association of males with blue and females with red. Whether this association is caused by innate gender differences or the result of cultural conditioning is not evident. The student researchers then administered the Big Five Personality Test, a questionnaire with 61 multiple-choice questions that purports to give a pretty accurate assessment of a person's personality. Interestingly, males showed no correlation of color preference with personality, but females did. Females who preferred red were calm and relaxed, females who preferred blue were conscientious, those who preferred green were agreeable, and those who preferred yellow were closed-minded.

This study, accomplished by undergraduates at the university, showed interesting outcomes while highlighting some of the procedural challenges that need to be addressed. First is the need for controls, which the students didn't offer in their study. It would have been important to control for all kinds of biases (time of viewing the colors; shape of the color samples; different brightness or hues of the same color, etc.) when they were assessing color preference. Second, the study relied heavily on a single personality test (the Big Five Test), which involves answers from subjects and could be flawed when translated into an emotional state. Finally, there was no attempt to tie the results back to physiology or neurobiology. These three simple problems are only the tip of the iceberg when doing studies like the association of color preference with personality or emotion.

One way to overcome these problems, especially the tie back to physiology, is to monitor the subjects of the study for physiological factors like heart rate and skin conductance, which are both indicators of heightened or aroused emotional state. Daniel Oberfeld and Lisa Wilms were interested in how the hue, saturation, and brightness of colors affected emotional response. They used these physiological indicators to explore the correlations of color hue, saturation, and brightness with emotional state by systematically altering these aspects of color. They were able to show clearly that all three color axes (hue, saturation, and brightness) were involved in subjects' emotional response to colors. Specifically, achromatic colors (white, gray, or black) resulted in deceleration of heart rate while chromatic colors (red, blue, green, yellow, etc.) resulted in accelerated heart rate. In addition, saturated and very bright colors resulted in higher skin conductance (more arousal) than other colors.

Some authors attempt to circumvent the problems of establishing an association between color and emotions by using high-powered statistical approaches. In one such study, associating colors with words for their emotional state was accomplished with multivariate statistical approaches. Other data were collected for the participants in the study, such as cultural conventions and personal experience. This approach is slightly better than the one taken by our undergraduates from the Philippines. It can order the different variables involved, of which there are many. For instance, hue is found to be the most important variable involved in the emotional response to color. The study established that cultural conventions and personal experience were involved but not as much as the initial emotional response to the hue of the colors.

Emotional state caused by color can have wide-reaching effects on people. Aseel A-Ayash and colleagues have looked at the impact of color on university students' reading capacity. By looking at heart rate, emotional response, and reading comprehension they were able to show that pale color conditions were associated with a relaxed, calm, and pleasant state. They also show that reading comprehension is better under bright hues. As with other studies, they could also show that blue increases the degree of relaxation and calmness of a participant. Whether these results can be used to enhance educational parameters, such as reading comprehension, remains to be seen.

One of the more precise studies of color correlation with emotion was done in 2016 by Avery N. Gilbert and colleagues. They used 194 participants who sat in front of a touch screen computer and were shown twenty words, one at a time, associated with emotions. They also were simultaneously seeing a color palette and were asked to select a color that "matched" the emotion word they were reading. It is clear that as with other color emotion correlation studies, color matching is dependent on gender and age. The statistical analysis in the paper is pretty strong, but they also figured out a way to visualize the color/emotion correlations. By creating a two-dimensional fourteen-by-fourteen block canvas (196 possible blocks), they were able to plot all 194 answers for the twenty emotion words presented to the participants. Imagine if all 194 participants saw the emotion word "angry" and all touched the hot red color from their color palette. Now imagine if the 194 participants saw the word "calm" and touched the cool blue button. The word emotions "anger" and "calm" experimentally now have visual characterization. The resulting two "canvases" would look very different though—one all hot red and the other cool blue. Kind of like the monochromatic paintings of Yves Klein but not all blue.

In reality, people do not agree with each other in this way though, and the canvases look more like they were created by Piet Mondrian using identically sized squares. In fact, Gilbert and colleagues call the figures "Mondrians" in their paper. Each emotion has about six to ten different colors that the 194 individuals choose for a given emotion word, so they are quite interesting to view. Just as you can enter an art gallery and easily discern a Klein from a Mondrian, so you can differentiate between an emotion angry Mondrian and an emotion calm Mondrian. The former has more red in it (a substantial number of black squares too), and the latter has a lot of cool blue squares in it (with some pink squares too). The individual opinions linking color to emotion are combined into a unique Mondrian for each emotion. And there are some discernable trends concerning association of color with emotion. Angry, tense, irritated, and anxious are all reddish Mondrians, but all distinct from each other. The amount of red in these four emotions tails off as one goes from angry => tense => irritated => anxious. Other emotion words have similar patterns. Sleepy, sad, bored, and tired have a lot of purple and black in their Mondrians, and energized, alert, happy, and healthy have a lot of green and yellow in theirs. But again each emotion produces a different Mondrian. Gilbert and colleagues suggest that the color emotion linkage influences how humans react to colors in advertising, and there is good evidence that this is a correct way to view the color associations, as we discuss below.

If one visits Adam Pazda's website, a bright red header hits you flat in the face. It's no wonder though, as Pazda is an expert in the emotional response to colors, red in particular. He has written papers with titles like "Processing the Word Red Can Influence Women's Perceptions of Men's Attractiveness" and "Women's Red Clothing Can Increase Mate Guarding from Their Male Partner." In these papers he uses clever approaches to test hypotheses about how the color red influences emotions in a sexually charged scenario. He clearly shows that women who wear red induce a much more extreme protection response from the males they associate with, and that the "lady in red" is perceived as a sexual threat to other women not in red. But his most interesting work to date involves how humans perceive the emotions of others via such facial coloration. Because the human face has an intricate network of blood vessels coursing through it, simple modulation of blood flow of the face can easily change color of the face. Perhaps because humans are very cued into faces for initial interpretation of social interaction, color of the face has become an important factor in addition to the expression on the face. Decoding emotions of other people is an important characteristic of humans in social groups. Some studies have shown that color of the face may be more important in deciphering the emotions of others in social situations. Pazda has clearly shown that the color of the face is more important than the color of other parts of the body in this social interaction; in addition, facial color can convey to a potential mate how healthy the potential mate might be. Red, as we discussed in chapter 7, is one of those colors that induces a wide range of responses in humans. Perhaps the most primitive response is in a sexual context.

Spence Accounts

Marketing assuredly has a great impact on our preferences for color. Charles Spence has made a brilliant and fun career out of looking at peoples' preferences for color, texture, sounds, and other sensory characteristics. His work on color in marketing and consumer preference for food is particularly interesting. His approach to examining the impact of color is quite simple. Offer subjects the same food under different conditions and watch what happens. There are two aspects to this kind of work with respect to color—the color of the food itself and the color of the serving device (the plate or cup or utensils). Some of the nuances of the color of food are obvious. You simply don't eat blue meat, nor do you drink green wine (although we have indulged in green beer on St. Patrick's Day). But Spence demonstrates other more refined preferences related to color of the food itself. Many of the preferences humans have with respect to color of food are culture specific. What is an appealing food color in one culture might be entirely different in another. Perhaps more interesting and easier for marketers to control is the color of the serving devices like plates and cups. For one of these experiments, published in the appropriately named journal *Flavour*, Spence and colleagues served sweet mousse on different colored plates of different shapes. It turns out that the color of the plate has a lot to do with the perception of sweetness, while shape is much less important. White and black plates were used and combined with different shaped plates. The shape had no impact on how sweet the participants perceived the mousse, but white plates made people perceive a much sweeter mousse.

Spence is also the researcher who did the famous red versus white Coke cans sweetness test. In 2011, Coca-Cola issued a white can in contrast to its normal red can. Spence tested whether the coke in the red can was sweeter than in the white can (even though the beverage was identical in both cans). In yet another experiment Spence tested whether the color of a coffee cup influenced peoples' perception of how tasty a hot drink is. Participants were served hot chocolate in four cups of different color (red, orange, white, and dark cream). The trick here is that only two of the cups were sweetened with sugar in different combinations. Orange and dark cream cups fared much better than the white or red cups with respect to the participants' perception of the sweetness of the beverage. On the other hand, the cup color had no impact on the perception of sweetness and chocolate taste and smell. These experiments have led Spence to say, "Color is the single most important product-intrinsic sensory cue when it comes to setting people's expectations regarding the likely taste and flavor of food and drink." Color is so important in this context because it is such a weak CSV, and hence it is labile within and across cultures. To show this concept we point to a series of interesting studies on consumerism. Luiz de Mello and Ricardo Pires Gonçalves, using choice experiments, showed that labels are incredibly important in how consumers view and buy wines. They offered a wide range of differently shaped labels with different colors on them. Surprisingly, black and square is the big winner. The choice has nothing to do with the quality of the wine, as all price categories of wine garner responses in the same way.

From the very first use of color in archaic human settlements, to cave drawings, to *Blue Boy,* and now to Charles Spence, color has played and continues to play a role in human communication and thought. Its biggest impact that we can easily unwind is how consumers react to color when purchasing items. And its biggest impact in the context of our development as a conscious species may be in art.

-isms

We have spent seven chapters describing what color is. But have we really? There is the overused philosophical question, Is the *red* you see the same *red* that I see? It's one of those philosophical questions many readers will have pondered over in philosophy class or recreationally in college, like Otter, Kate, and Boone in *Animal House* when they meditate over the possibility that the solar system we live in is no more than an atom in the finger of some infinitesimal giant and so on. We know that color in the context of the physical world is one thing. But it morphs into something completely different when we examine the organismal world. What we think we have nailed down in the organismal world then quickly transforms into something very different when we consider how color has influenced our social and cultural existence. Color is a very slippery concept indeed (by the way, Otter, Kate, and Boone's universe as atoms in a fingertip concept only made sense in their altered state). In this chapter we ask about color once again and what color is. But we ask this question in a different, more philosophical, way.

We are of course talking about the philosophical realm whenever we pose a problem like, Is the *red* you see the same *red* I see? Color has been at the heart of philosophical discussion and development for as long as philosophers quit their day jobs and started earning their living from their thinking. It is because color perception is such an interesting topic in a philosophical context AND we know a lot about it scientifically that it poses a unique problem in philosophy and in the development of our consciousness. Compare color to other philosophically focused topics like "God" or "existence," where we really have little if any scientific background to work with. Color is almost the best test case for the utility of philosophy as a discipline. If we can't discuss and perhaps even solve it in a philosophical context, then forget about God or existentialism.

The 18th-century polyglot David Hume sums the philosophical problem up with this simple quote: "Sounds, colors, heat and cold, according to modern philosophy are not qualities in objects, but perceptions in the mind." Philosophers in general want to know whether the properties of objects (things that we see, smell, taste, feel, and hear) are among the properties they are agreed to have in reality. The critical concept here is "in reality."

We never really know what is real; we can only eliminate things that aren't real from our lexicon or ontology of the natural world. Consider that we use the word "species" in biology all the time. A species can be defined in many different ways, and indeed there are dozens of species concepts and definitions that biologists use in plying their trade. Some of these definitions and concepts are more objective than others and hence are considered superior to the less objective ones, but all concepts and definitions rely on what in reality is a species, and we never really can know that. Evolutionary biologists can guess, statistically speculate, or come together and agree about what a species is, but that's all.

The word "color" is like the word "species," and the arguments about what a species is are as abundant as the arguments about what color is. We think we know what these things are, but we never will "in reality" know. Hence, philosophers feel compelled to explore all and any alternatives to what we as humans might feel is the reality of the situation. This has led to a lot of terms being bandied about in the context of color and in seeking a rational, philosophically sound perception of what color means to us humans. If anything, as M. Chirimuuta proclaims, "color hovers uneasily between the subjective world of sensation and the objective world of fact."

The terminology of color philosophy is populated with a forest of "isms." There are a large number of terms, some similar to each other and others very different. It is difficult to see through this philosophical forest, and it can become a dark place indeed if you wander off the path. Table 8.1 shows a small list of -isms we ran across in looking at the topic of color and its philosophical underpinnings. We are sure there are more out there, but these were the most easily uncovered. The table also gives definitions of the terms in the context of color. You might ask, Do we really need this many terms to describe color? Indeed, we asked the same question, but it turns out that the precision with which philosophers have examined color has led to the efflux of terminology when discussing color.

Table 8.1. A (not exhaustive) glossary of color-related -isms in no particular order

Realism: physical phenomena are physical properties of objects.
Color realism is when colors are physical properties of objects; what best explains why objects *look* colored is that they *are* colored.
Color antirealism agrees with the premises of color realism (that spectral power distributions and color experiences are jointly sufficient to explain the gamut of chromatic phenomena) but says that realism is not needed in our explanation of color (i.e., commonsense realism [see next entry] is not a requirement for a description of color).
Commonsense realism denies abstractions and emphasizes the ability of any individual to perceive the nature of the world directly.
Anthropocentric realism is based on inquiry into colors as experienced by a single species, namely humans. Since variations across species present serious problems for color realism, most color realists accept anthropocentric realism (they have to).
Quality realism is where colors are the result of the qualities of the surfaces of the objects around us rather than mysterious neural impulses inside the brain.
Subjectivism: where colors are *dispositions* to cause color experiences.
Eliminativism: suggests that despite objects' appearances, they are not actually colored; fictionalism (next entry) is a form of elimativism.

Fictionalism
Color fictionalism comes in two varieties: descriptive and prescriptive.
Prescriptive color fictionalism assumes that color uses a statement that is false, and recommends that we carry on employing the discourse with the statement, as if this were not the case. It is one of the two color fictionalisms and a way of thinking that allows discourse about colors without colors being real.
Descriptive fictionalism is where color is treated as a fiction from the outset, or color is treated as a pretense. As Dimitria Electra Gatzia implies, the descriptive fictionalist is merely pretending that objects are colored in order to continue the discourse on color.
Neo-conservative fictionalism is not the concept that prompted Donald Rumsfeld* to utter, "Known knowns; things we know we know. Known unknowns; we know there are some things we do not know. Unknown unknowns; things we don't know we don't know" statement. This describes a way of thinking that says there are theories superior to the obvious. The theory may not be accessible to us but accessible to creatures with superior cognitive capacities than ours.

---

* The actual quote from Donald Rumsfeld was "Reports that say that something hasn't happened are always interesting to me, because as we know, there are known knowns; there are things we know we know. We also know there are known unknowns; that is to say we know there are some things we do not know. But there are also unknown unknowns—the ones we don't know we don't know. And if one looks throughout the history of our country and other free countries, it is the latter category that tend to be the difficult ones."

Dispositionalism: where colors are identified with psychological dispositions. For example: blueness = the disposition to look blue to normal perceivers in normal conditions. The perception of that very color will change with the disposition of the observer.

Primitivism: suggests that colors are simple, intrinsic properties, which exist as a characteristic of the surfaces of material bodies.

Physicalism: where colors arise from and are identical to physical properties of an object, which are physically relevant reflectance properties of the surface of a material body.

Organism: most color -isms would not exist without organisms with color sensing. So, organisms are ultimately at the heart of color -isms and why we include this -ism in the table.

Color adverbalism: a term invented by M. Chirimuuta. She described this kind of color theory to overcome the excesses of objective analysis of our senses. For instance, if we experience red, whether in ordinary visual instances, in a dream, or in a hallucination, it involves a mental act through which the object being viewed becomes acquainted with red. The object isn't physically or realistically associated with red, but rather is associated with red like an adverb is associated with a verb.

Specific color philosophy discussion schemes differ from each other, but the best way to view the problem is to accept a general three-way schism in terminology that hinges on realism, subjectivism, and eliminatavism arguments. Some philosophies of color divide the argument into four categories: eliminativism, dispositionalism, primitivism, and physicalism, which to us is just a more fine-grained version of realism versus fictionalism. We suppose that as a physical and a biological scientist we are both preadapted to being color realists. This is because the realms of physics and biology (neurobiology especially) have contributed so much to our understanding of color in nature. What we have learned in these domains appears very real to us. We might also see the easiest way to view the argument is as a two-way split between realism and antirealism, or physicalism and fictionalism. Subjectivism is an interesting alternative to the two extremes of realism and eliminativism and sits in between the two. Also, as scientists our ideas live and die by how they stand up to scientific testing. Such tests strive to explain the strange things we see in nature, and color is one of the strangest physical phenomena out there. We can only eliminate possibilities through scientific endeavor. The more things we can eliminate, the closer we get to the best explanation of a natural phenomenon. Science strives to have the greatest explanatory power, though, and this is why we say we "suppose" we are realists or, more precisely, antifictionalists. We fall to the side of antifictionalism because we have a great deal of faith in the scientific process. Another way to put this is to say it is tempting to simply fall on a dichotomous way of describing the problem as between realism and fictional (elimitivism)—color is either real or it is not. But because there are some philosophers who cannot stomach either of these two extremes, subjectivism (and, if you want to go there, dispositionalism and primitivism) is a somewhat appealing way to think about color. And as we will soon see, having been trained as a scientist does not necessarily mean one has to be a color realist.

If all of this doesn't sound confusing, then you are more than likely someone who has thought about the philosophical problems a lot or someone who didn't need Table 8.1 to navigate what we have said so far. We have mixed some of the terms together, which might drive our philosopher friends a little batty, but it is the best way we can think of that might make explaining this easier. One of the clearer expositions of color realism appears in a review of the subject written by Alex Byrne and David R. Hilbert entitled "Color Realism and Color Science," published in 2003 in *Behavioral and Brain Sciences*. Well, maybe not so clear, as thirty-two separate responses to the original review were later published in the journal. So, we guess if you didn't find the discussion above confusing, you might be an author of one of those thirty-two responses.

No one likes to be criticized, right? Wrong! In science, and apparently especially in color realism and color science, being criticized is a badge of honor. Why? Because by being criticized, the question you originally addressed becomes clearer to your readers and clearer to you. You advance the philosophical argument and the world is better for it. And you get to publish a response to all of the criticisms, which Byrne and Hilbert did quite eloquently. They fall on the side of color realism and go through the various philosophical arguments debunking opponents of color realism (which of course those opponents responded to in turn). They favor a color realism that wants colors to be physical properties of objects, residing in objects that reflect or transmit light. In other words, reflectance of light embodies the reality of color. To us there were several important outcomes of this rather interesting interchange of minds (all forty-two of the minds involved including Byrne and Hilbert; see Appendix). Actually, not all of the minds want to debunk realism, but rather some realists disagree with Byrne and Hilbert's account of it.

The first essential thing to get is that tomatoes can be used for more than just food (and in some cases, strawberries or radishes too). Understanding the redness of a tomato appears to be one of the more important tasks of the whole realism/antirealism argument, and examples using tomatoes abound in both the original article, the responses to the article, and the literature in general. The antirealist tomato argument goes like this: a tomato has important physical aspects with respect to its color. Its surface reflects physical light and has other microphysical, biological, and chemical properties. It affects perceivers of the tomato in certain ways. There are no other properties that the surface of the tomato needs to have to explain a human's visual interaction with a tomato. The color, or rather alleged, color of the tomato does not help us to causally explain our visual reaction to the tomato. But wait: if this is true, then no perceptible physical property of the tomato can help us understand the color of the tomato. If this last dictum is true, then there is no logical reason to assume that the tomato has any physical color.

We also need to add another -ism to Table 8.1—metamerism. This concept is a lot like one we have discussed throughout this book—homology. Metamerism refers to the fact that different reflectances from objects can in fact deliver the same information about light to the brain. Metamerism occurs because of the many different ways the three kinds of cone cells can deliver information to the brain. It is reasonable to assume that there are many ways stimuli from the cone cells can combine to converge on each other and deliver the same or nearly the same signal for color to the region of the brain where color information is processed. If metamers exist, then it is difficult to buy the reflectance realism of Byrne and Hilbert, because no single combination is deterministic of a color: "Determinate colors cannot be identified with specific reflectances because there will typically be (indefinitely) many reflectances that result in the appearance of a given determinate color and no motivation for choosing between them."

Another problem involves the broad variation among human populations with regard to color vision. In studies of how people process green hues, there is a range of 490 to 520 nm where subjects will identify a unique green. That is, one individual might identify light of a wavelength of 495 nm as true green. A second individual might identify light of wavelength 510 nm as true green. Given that 15 nm on either side of any individual's "true green" on average will look bluish on one end and yellowish on the other, this range from 490 to 520 is quite large. So, in this case, a reflectance cannot be both yellowish (to one individual) and bluish (to a second individual) and be a physical property of the object. Byrne and Hilbert discard this argument by pointing out that "the conclusion is not that people rarely see objects as having the colors they actually have, but that they rarely see objects as having the *determinate* colors they actually have." There is a difference between determinate color and actual color.

Which brings us to a well-known thought experiment first conceived by John Locke, the great philosopher of the 17th century, called the "inverted spectrum." Locke's thought experiment created two kinds of people on the planet; those who see colors the normal way (Nonverts) and those who have an inverted perception of colors (Inverts). This Nonvert/Invert world has been suggested to be anathema to realism and physicalism. Any random pair of people could be seeing the world in completely different ways (if the pair were a Nonvert/Invert pair) and hence the reflectance of an object has little to do with the perceived color. Some of the Nonvert/Invert argument can be easily pushed aside as irrelevant to physicalism. But one specific Nonvert/Invert world is actually quite interesting. It is the world where both the Invert and Nonvert call blood red and broccoli green. When both of these kinds of people are looking at a tomato (yes, the tomato returns), they both see it as red, but they are experiencing the tomato in completely different ways. Physically it is not the same experience for the Nonvert as it is for the Invert. In this Nonvert/Invert world it is pretty easy to suggest that reflectance physicalism is not what color is about. An Invert or a Nonvert who grew up with a continually shifting random assignment of colors to objects would essentially never get what red is even though the physical reflectance of red light is processed by their eyes.

Inverted spectra scenarios that involve Nonverts and Inverts have not been observed in humans, but science writer Vi Hart pointed out in 2016 that virtual reality (VR) might be the key to creating true analyzable Nonvert/Invert worlds. VR can transform any light into any Invert scheme and create the Inverts needed to do the Nonvert/Invert comparison for real. Hence it also could lead to tests of Byrne and Hillbert's major ideas about color physicalism.

Color scientist C. Larry Hardin recognized that the validity of color physicalism relies on having a unique red (unique any color, for that matter). That is, a red that can be recognized as such by researchers specifically and people in general. This red can be empirically determined using color matching experiments where subjects are given the name of a color and then challenged with a palette of color chips to match a chip with the name. Hardin suggests that "we might, for example, decree that the most frequently chosen chip is to be unique green (red)." Like the typologists of Aristotle's time, there is a need to reduce the noise in red things to an essence of red. But here it is a very precise highly statistical concept of the essence of unique red. Hardin recognizes the subjectivity of the approach, as he also says, "but we could decide otherwise." Hardin does not agree with color physicalism for several reasons, one of which is that color physicalism cannot completely and objectively solve the unique red problem.

On the other hand, Hart implies the critical ingredient to color physicalism and realism hinges on the public context-dependent formulation of unique colors. In this realm, colors become statistical concepts; an average that can be assessed as robust or not so robust. VR offers a great comparative experimental approach to pinning down the statistical significance. However, VR cannot answer the question with impunity. Rather, VR will allow us to create data sets that will give us a statistical sense of unique red and allow color physicalism to fit in.

We faced a similar problem earlier in this book when we discussed how Charles Darwin changed our view of the world by pushing evolutionary thinking away from essentialism and typological methods and more toward a populational way of thinking where the variation in a system was the thing to watch. As we described in chapter 4, without this shift in his thinking, he would not have gotten natural selection. Perhaps the whole color problem boils down to a bigger respect for variation in color perception. We might learn more about how our brains work by going down a road where the essential concept of unique colors is not so prevalent in philosophical treatments of color. But we also point out that one branch of evolutionary biology has been and remains very typological—systematics. Systematics and other less typological and even non-typological approaches to evolutionary biology exist fairly harmoniously within that realm of science. We make this distinction as a caveat to complete removal of physicalism, especially reflectance physicalism, from color science. It may have its role, after all, in a better understanding of what happens between the time light hits our retinas and the moment we perceive color in our noggins.

Color Leads to Hard/Soft Problems in Consciousness?

When color is processed by our brains, how does it affect human consciousness, and our understanding of it? The philosopher Nigel Thomas once asked, "Does it matter to consciousness researchers whether colors are 'really' on the surfaces of objects or 'really' in the mind?" He wondered about the relevance of color philosophy to consciousness, and it is this question that is the main link to how color philosophy will impact how we understand consciousness. The problem of the human mind is a huge one. Thousands of books, millions, if not billions, of words, and much human thought have gone into trying to understand the human mind. The approach to color that we have taken in this book has been very physical and biological, addressing what neuroscientists call the easy problems of the mind. Easy problems include figuring out where our perceptions of the outside world originate in the brain and how this perception works. In his book *The Astonishing Hypothesis*, Francis Crick eloquently wrote that "a vast assembly of nerve cells and their associated molecules" are responsible for the mind and the emergence of consciousness. The easy problems deal with those physical aspects of perception caused by nerve cells and molecules. However, the holy grail is in the realm of what neuroscientists call the hard problem of consciousness. This problem is indeed hard, as any answer to it wants to link an emergent property of our neurobiology (mind) with physical, molecular, and chemical information. The tough spot is that we need the easy problems solved to shore up any ideas we have about the hard problem, while the answers to the easy problems don't get us all the way there.

It is straightforward to reconstruct some of the evolutionary history of sensory processing for color in our species and other organisms. For instance, an example can be seen in a book Rob DeSalle and Ian Tattersall published in 2012. Using Antonio Damasio's ideas about emotions, they reconstructed the evolutionary history of emotion in animals. This reconstruction, not surprisingly, suggests that our species is quite unique in how we deal with the outside world emotionally. What is surprising in this reconstruction is the stepwise fashion in which we humans came to our emotional systems today. Likewise, we show here that color perception arose in a stepwise fashion with some lineages using color perception in unique ways and others using it other unique ways. What we haven't shown is how color perception evolved into consciousness. This is the hard problem. Color is a beautiful, engaging, and practical tool for delving into this hard problem. Who knows, maybe color will be the key to unlocking the mystery of our consciousness.

EPILOGUE

The Color of Existence

We have traversed a spectrum of topics on color. This spectrum has led us through the physics and chemistry of the early universe where the photons responsible for our color perception started, to making a case for color being the answer to explaining our consciousness. Photons are all around us and the most significant information they deliver to our planet is in the guise of wavelengths. The wavelengths of photons are what initiate the perceptions we have of color. They are physical entities that our visual system collects and processes physically, biochemically, and neurologically. Our biology and the biology of most of the organisms on the planet have evolved to detect these wavelengths for a multitude of purposes. Plants, algae, and some bacteria use this stimulus for food. Animals use it for information. Humans have used it for art and fashion. The evolutionary mechanisms that steadied our human visual system and color perception capacity is also discussed at length in this book, and because eyes are so important to such perception we have spent some time on how eyes work in general and how our eyes work specifically.

Color perception is tricky business. To perceive colors in a human way, a fully functional retina is not necessarily needed. Witness Neil Harbisson, a self-described monochromat cyborg. Harbisson was born without the right opsins in his retinal cone cells to perceive color, so the world through his eyes is black, white, and shades of grey. He has gotten around this inborn problem though through the cyborg in him. He is outfitted with a cameralike device near his forehead and wired to the back of his head and eventually to his brain. When he points the camera on his forehead at something with color, the camera picks up the color, processes it and makes sounds that in turn make vibrations. So, for instance, dirty yellow socks will make a high-pitched sound and a lower sound is made when the camera detects a red handkerchief. So what do the sounds mean to his brain? Just like a blind person can use the sense of touch to "see" letters, Mr. Harbisson uses the vibrations from the sounds to "see" color. He has, in other words, trained himself to interpret the vibrations as colors. His green is nowhere near what our green is but Harbisson knows what green is and what the other colors are in his own mind.

In a similar way some organisms on this planet have evolved organs that help them to "see" without their eyes. We have already discussed the viper sensory pit that allows certain snakes to perceive objects that emit infrared radiation. The pit viper is in essence seeing in a whole new wavelength range without the use of its eyes. Other organisms have learned to use electrical fields to locate prey, such as the duck-billed platypus, which has an intricately evolved organ in its bill that senses the electric fields of other organisms fairly precisely. It has eyes, but they are not used to locate prey and food items. The star nosed mole uses several tiny appendages coming off of its nose to feel around its immediate surroundings for prey items. In this way it can "see" miniscule nematodes, locate them quickly, and devour them. And don't forget that we humans have artificially created a rainbow of colors from the RGB system; colors that look real, but are nothing more than the combination of red, green, and blue.

Color has grandly influenced our human behavior and sociality. We are a species driven by color in many ways and our cultural evolution oftentimes involves color as a driving force or a cultural survival vehicle. The actual colors of human hair, eyes, and skin are wonderfully variable but unfortunately perverse unscientific ideas about this variation has led to some of the worst human tragedies and behavior we know of.

The evolutionary view of how our senses evolved indicates that we are in many ways very unique in how the sensory information is processed in our brains. It is also evident from Ian Tattersall's writing on human consciousness that language and all of the very human things we do around language are most important developments in the emergence of the mind in our species. As Tattersall so eloquently puts it, we are the only species on this planet that can think about thinking and in doing so create "mental constructs of alternative versions of the world." The brain of our species, *H. sapiens,* was adequately neurologically wired 200,000 to 300,000 years ago to think symbolically, yet the best evidence we have is that we only started to think symbolically about 100,000 years ago. Tattersall suggests that there was a purely cultural trigger to our symbolic logic, and that was the invention of language by small populations of *H. sapiens* living in Africa about 100,000 years ago.

It is not surprising then, that philosophical questions around consciousness are very much about language and symbolism. Perhaps language is the key to inserting color perception as a link to consciousness. Whereas we may still have some way to travel to unlock the hard problem of human consciousness using color, much of the easy problem is unraveling before us as a result of modern neurobiology, genetics, and evolutionary research. We have tried to take an evolutionary approach to understanding color and the easy problems of the how color works and what color is in this book. We hope this approach has been illuminating. Because our perception of color practically has no limits though it is critical to keep this in mind when using color to unravel consciousness. Understanding color, combined with language, may very well be the key to understanding the hard problem of the mind.

Appendix

Author	Summary	real or antireal	Life Focus
Edward Wilson Averill	agreement on what is "red" is important	realism*	Philosopher
Aaron Ben-Ze'ev	colors are not physical properties	antirealism**	Philosopher
Michael H. Brill	subjectivism is unwieldy	realism	Physicist
James J. Clark	color determined by physical/ecological laws	realism	Computer Scientist
Jonathan Cohen	need to relativize colors to perceivers	realism	Philosopher
Frans W. Cornelissen, Eli Brenner and Jeroen Smeets	true color exists in the eye of the observer	antirealism	Neuroscientist
Lieven Decock and Jaap van Brakel	there is no all-embracing ontology for color	realism	Philosophers

Don Dedrick	philosophical and the empirical are not so disparate	realism	Philosopher
Brian V. Funt	how to amend the definition of reflectance-type	realism	Computer scientist
Martin Hahn	metamerism does not pose a problem for realists	realism	Philosopher
Stephen Handel and Molly Erickson	match supports the physicalism of color and timbre	realism	Psychologists
C. L. Hardin	no principled way of determining when we see colors as they really are	antirealism	Philosopher
Scott Huettel, Thomas Polger, and Michael Ri.. y	dispositional account is promising if understood in an ecological framework	realism	Phil, Neurosc, Psych
Frank Jackson	not plausible to hold that color experiences represent things as having the physical properties	realism	Philosopher
Rolf G. Kuehni	make a clearer case for reflectance types	antirealism	Textile chemist
John Kulvicki	the full intrinsic nature of colors is revealed to us by color experiences is false	antirealism	Philosopher
Bruce J. MacLennan	more realistic to embrace color's full phenomenology	realism	Computer Scientist
Laurence T. Maloney	correspondence between perceived surface color and specific surface properties.	realism	Psychologist
Mohan Matthen	categories capture objective truths about the environment	realism	Philosopher

Barry Maund	color realism concerns color language or color concepts	antirealism	Philosopher
Rainer Mausfeld and Reinhard Niederée	the role that "color" plays within perceptual architecture, and the complex coupling to the "external world" needs clarity	antirealism	Psychologists
Erik Myin	antirealist intuitions that flow from the specificity of color *experience*	antirealism	Philosopher
Romi Nijhawan	"perceived" position and color are not properties of "real" objects	antirealism	Computer Scientist
Adam Pautz	there are no physicalistically acceptable candidates to be the hue-magnitudes	antirealism	Philosopher
Adam Reeves	color is a factor analytic approximation to Nature	antirealism	Psychologist
Barbara Saunders	cannot substitute data sets for the life-world	antirealism	Anthropologist
Davida Y. Teller	argument for color realism collapses to an uninteresting terminological dispute	antirealism	Color Scientist
Dejan Todorovic	color illusions challenge color realism, because they involve a one-to-many reflectance- to-color mapping	antirealism	Psychologist
Robert Van Gulick	overly focused on input conditions and distal causes	antirealism	Philosopher
Richard M. Warren	both realism and antirealism are correct context is important	realistic antirealism	Psychologist

Zoltán Jakaba and Brian McLaughlin	commitment to color physicalism is a reason to reject realism	antirealism	Psychologists
Michael E. Rudd	no simple correspondence between experienced color and a stable class of physical reflectances	antirealism	Psychologist

* While realism and physicalism are not entirely equal, we refer to it here as linked to physicalism

** Also referred to as elminativism.

Phil, Neurosc, Psych = a Philosopher, a Neuroscientist, and a Psychologist

Appendix Legend. Responses to Byrne and Hilbert's "Color realism and color science." We have inferred from our naive reading of the responses the summary statement and the lean toward realism or antirealism. Sometimes our interpretation of antirealism could be argued to be subjectivism, and sometimes our interpretation of realism could be argued to be a less extreme form of physicalism. We apologize in advance for any misinterpretation to the authors of the thirty-two responses but hope the general trends of the responses are evident.

Bibliography

1. THE COLOR OF THE UNIVERSE

Understanding many of the concepts in this chapter can be shored up by reading Darwin's *On the Origin of Species*. Aspects of the evolution of the universe and Stephen Hawking's contribution to our understanding of it can be found in his book *A Brief History of Time*. The original no-boundary proposal of Hartle and Hawking was published in 1983, but many readers may find this paper a bit dense. Fortunately, Hawking was also a vivid communicator of science, and his *Universe in a Nutshell* discusses the no-boundary concept in lay terms. That last interview of Steven hawking can be found online at https://www.startalkradio.net/show/universe-beyond-stephen-hawking/. Andersen et al. (2019) clearly describes the brightness of the Big Bang in a fairly technical paper. Journalist George Johnson's description of the ten most beautiful experiments is best found in his book of the same name published in 2008. Darwin-Mendel communication is documented in the paper by Bizzo and Charbel (2009). Olshansky, Carnes, and Butler (2001) describe the knee joint and aging, and the classic Lewontin and Gould spandrels paper, which has been cited nearly ten thousand times, can be found as listed below. Philip K. Dick's short story is easily accessed in any one of the many collections of his short stories, such as the one listed below.

Andersen, Christopher, Charlotte Amalie Rosenstroem, and Oleg Ruchayskiy. "How Bright Was the Big Bang?" *American Journal of Physics* 87, no. 5 (2019): 395–400.

Bizzo, Nelio, and Charbel N. El-Hani. "Darwin and Mendel: Evolution and Genetics." *Journal of Biological Education* 43, no. 3 (2009): 108–114.

Darwin, Charles. *On the Origin of Species*. 1859. London: Routledge, 2004.

Dick, Philip K. *The Philip K. Dick Reader*. New York: Citadel Press, 1997.

Gould, Stephen Jay, and Richard C. Lewontin. "The Spandrels of San Marco and the Panglossian Paradigm: A Critique of the Adaptationist Programme." *Proceedings of the Royal Society of London. Series B. Biological Sciences* 205, no. 1161 (1979): 581–598.

Hartle, James B., and Stephen W. Hawking. "Wave Function of the Universe." *Physical Review D* 28, no. 12 (1983): 2960.

Hawking, Stephen, and Michael Jackson. *A Brief History of Time*. Beverly Hills, CA: Dove Audio, 1993.

Hawking, Stephen. *The Universe in a Nutshell*. Odile Jacob, 2001.

Johnson, George. *The Ten Most Beautiful Experiments*. Visalia, CA: Vintage, 2008.

Olshansky, S. Jay, Bruce A. Carnes, and Robert N. Butler. "If Humans Were Built to Last." *Scientific American* 284, no. 3 (2001): 50–55.

2. COLOR WITHOUT EYES

The strange idea that plants have minds is developed by Monica Gagliano in a paper written in 2017. Charles Darwin's ideas about plants can be found in his book *The Power of Movement in Plants*. Trail et al. (2011) discuss the Hadean oxidation state. Zuckerkandl and Pauling's classic 1965 molecular evolution paper is referenced below. Halobacterial rhodopsins are described in Bibikov et al. (1993), and Lynn Margulis's multiply rejected but finally published paper on endosymbiosis can be found in the *Journal of Theoretical Biology*. Shinichiro Maruyama and Eunsoo Kim describe algal evolution, while Eunsoo Kim explains endosymbiosis and the

origin of chloroplasts in this science blog https://blogs.scientificamerican.com/
artful-amoeba/green-alga-found-to-prey-on-bacteria-bolstering-endosymbiotic-
theory/?redirect=1. Daniel Trembly MacDougal's century-old paper on light and
development is also listed below. Varoqueaux et al. (2018), Oakley and Speiser
(2015), and Garm et al. (2011) discuss elements of eye in all kinds of squishy
animals. Finally, for a thrilling take on modern molecular biology and organismal
evolution, see Sean B. Carroll's 2008 book *The Making of the Fittest.*

Bibikov, Sergei I., Ruslan N. Grishanin, Andrey D. Kaulen, Wolfgang Marwan,
  Dieter Oesterhelt, and Vladimir P. Skulachev. "Bacteriorhodopsin Is
  Involved in Halobacterial Photoreception." *Proceedings of the National
  Academy of Sciences* 90, no. 20 (1993): 9446–9450.

Carroll, Sean B. *The Making of the Fittest.* London: Quercus, 2008.

Darwin, Charles. *The Power of Movement in Plants.* Appleton, MN: Appleton,
  1897.

Gagliano, Monica. "The Mind of Plants: Thinking the Unthinkable."
  Communicative & Integrative Biology 10, no. 2 (2017): 38427.

Garm, Anders, Magnus Oskarsson, and Dan-Eric Nilsson. "Box Jellyfish Use
  Terrestrial Visual Cues for Navigation." *Current Biology* 21, no. 9 (2011):
  798–803.

MacDougal, Daniel Trembly. *The Influence of Light and Darkness upon Growth
  and Development.* Vol. 2. New York Botanical Garden, 1903.

Margulis, Lynn. "On the Origin of Mitosing Cells." *Journal of Theoretical
  Biology* 14, no. 3 (1967): 225–274.

Maruyama, Shinichiro, and Eunsoo Kim. "A Modern Descendant of Early
  Green Algal Phagotrophs." *Current Biology* 23, no. 12 (2013): 1081–1084.

Oakley, Todd H., and Daniel I. Speiser. "How Complexity Originates: The
  Evolution of Animal Eyes." *Annual Review of Ecology, Evolution, and
  Systematics* 46 (2015): 237–260.

Trail, Dustin, E. Bruce Watson, and Nicholas D. Tailby. "The Oxidation State
  of Hadean Magmas and Implications for Early Earth's Atmosphere." *Nature*
  480, no. 7375 (2011): 79–82.

Varoqueaux, Frederique, Elizabeth A. Williams, Susie Grandemange, Luca
  Truscello, Kai Kamm, Bernd Schierwater, Gaspar Jekely, and Dirk Fasshauer.

"High Cell Diversity and Complex Peptidergic Signaling Underlie Placozoan Behavior." *Current Biology* 28, no. 21 (2018): 3495–3501.

Zuckerkandl, Emile, and Linus Pauling. "Molecules as Documents of Evolutionary History." *Journal of Theoretical Biology* 8, no. 2 (1965): 357–366.

3. COLOR WITH EYES

The references for this chapter start with Gavin de Beer's 1971 essay on homology. Walter Ghering and Kazuho Ikeo wrote a neat review of the eye master switch gene in 1999. Strausfeld et al. (2016) describe in detail the arthropod eye and its divergence and convergence. The phototactic system of *Platynereis dumerilii* is described by Jékely et al. (2008). Papers by Connor (2000), Turner (2014), and Gregory (1968) all address the issue of how the brain processes the information on color from our eyes. The book by Gibson (1950) lays out much of the basic biology behind color perception at the neurological level. The basics of color equivalency are discussed in detail in Mitchell and Rushton (1971). We also mention Gary Marcus' 2009 book *Kluge* in the text as a model for how the human brain has evolved.

Connor, Charles E. "Visual Perception: Monkeys See Things Our Way." *Current Biology* 10, no. 22 (2000): R836-R838.

De Beer, Sir Gavin. *Homology, an unsolved problem*. Vol. 11. Oxford: Oxford University Press, 1971.

Gehring, Walter J., and Kazuho Ikeo. "Pax 6: Mastering Eye Morphogenesis and Eye Evolution." *Trends in Genetics* 15, no. 9 (1999): 371–377.

Gibson, James J. *The Perception of the Visual World*. Boston: Houghton Mifflin, (1950).

Gregory, Richard L. "Visual Illusions." *Scientific American* 219, no. 5 (1968): 66–79.

Jékely, Gáspár, Julien Colombelli, Harald Hausen, Keren Guy, Ernst Stelzer, François Nédélec, and Detlev Arendt. "Mechanism of Phototaxis in Marine Zooplankton." *Nature* 456, no. 7220 (2008): 395–401.

Marcus, Gary. *Kluge: The Haphazard Evolution of the Human Mind*. Boston: Houghton Mifflin Harcourt, 2009.

Mitchell, D. E., and W. A. H. Rushton. "Visual Pigments in Dichromats." *Vision Research* 11, no. 10 (1971): 1033–1043.

Strausfeld, Nicholas J., Xiaoya Ma, Gregory D. Edgecombe, Richard A. Fortey, Michael F. Land, Yu Liu, Peiyun Cong, and Xianguang Hou. "Arthropod Eyes: The Early Cambrian Fossil Record and Divergent Evolution of Visual Systems." *Arthropod Structure & Development* 45, no. 2 (2016): 152–172.

Turner, R. Steven. *In the Eye's Mind: Vision and the Helmholtz-Hering Controversy*. Princeton, NJ: Princeton University Press, 2014.

4. THE COLORS OF EVOLUTION

Tim Caro's historical treatment of color and Wallace's place in that history can be found in Caro's 2017 review article. The nomenclature of colors that Darwin used on the voyage of the *Beagle* is referenced below too. Wallace's *Letters from the Malay Archipelago* and Darwin's *The Voyage of HMS* Beagle can be consulted for a taste of the wonderful writing of these two giants. Jerry Coyne's description of the Kettlewell problem is described on his blog *Why Evolution is True* (https://whyevolutionistrue.wordpress.com/) and in his 2002 *Nature* paper. The original work by Nachman and colleagues is listed so that the reader can peruse a first-class research paper. The capacity for human NIR vision is in the publication by Palczewska et al. (2014). Descriptions of unconventional color vision by Marshall and Arikawa (2014) and Musilova et al. (2019) are also referenced below.

Caro, Tim. "Wallace on Coloration: Contemporary Perspective and Unresolved Insights." *Trends in Ecology & Evolution* 32, no. 1 (2017): 23–30.

Coyne, Jerry A. "Evolution Under Pressure." *Nature* 418, no. 6893 (2002): 19–20.

Darwin, Charles. *The voyage of HMS* Beagle. New York: Penguin Random House, 1910.

Marshall, Justin, and Kentaro Arikawa. "Unconventional Colour Vision." *Current Biology* 24, no. 24 (2014): R1150–R1154.

Musilova, Zuzana, Fabio Cortesi, Michael Matschiner, Wayne I. L. Davies, Jagdish Suresh Patel, Sara M. Stieb, Fanny de Busserolles, et al. "Vision Using Multiple Distinct Rod Opsins in Deep-Sea Fishes." *Science* 364, no. 6440 (2019): 588–592.

Nachman, Michael W., Hopi E. Hoekstra, and Susan L. D'Agostino. "The Genetic Basis of Adaptive Melanism in Pocket Mice." *Proceedings of the National Academy of Sciences* 100, no. 9 (2003): 5268–5273.

Palczewska, Grazyna, Frans Vinberg, Patrycjusz Stremplewski, Martin P. Bircher, David Salom, Katarzyna Komar, Jianye Zhang, et al. "Human Infrared Vision Is Triggered by Two-Photon Chromophore Isomerization." *Proceedings of the National Academy of Sciences* 111, no. 50 (2014): E5445–E5454.

Syme, Patrick. *Werner's Nomenclature of Colours*. Edinburg, England: James Ballantyne and Co (1814).

Wallace, Alfred Russel. *Alfred Russel Wallace: Letters from the Malay Archipelago*. Oxford: Oxford University Press, 2013.

5. GARY LARSON TO THE RESCUE

This chapter relies on a lot of references, starting with the Cuthill et al. (2017) review article in *Science* magazine. The role of ecological chemistry in mimicry can be found in Brower (1969), and the use of *Daphnia* by Arenas et al. (2015) to study aposematism is also included in this list. Alfred Russel Wallace's wonderful book *Mimicry, and Other Protective Resemblances Among Animals* was the first document to outline a research program using color in nature. Coral snake mimicry is discussed in Greene and McDiarmid (1981), Ackali and Pfennig (2017), and Rabosky et al. (2016). Paul Ehrlich and Peter Raven's classic paper in *Evolution* is a good place to start for understanding "arms races" in nature. Nijhout (1978), Westerman et al. (2018), Kronforst and Papa (2015), as well as The Heliconius Genome Consortium (2012), describe the *Heliconius* wing pigmentation work discussed in this chapter. Stoddard and Prum (2011), Gilbert (2015), and Dale et al. (2015) all discuss the ins and outs of bird plumage and coloration. Stevens and Merilaita (2008) look closely at camouflage in their article on the subject. Pieribone, Gruber, and Nasar's 2005 book, *Aglow in the Dark*, is a wonderful introduction to fluorescence and its diversity in the marine realm. Iridescence, fluorescence, and structural color are discussed in Protzel et al. (2018), Marshall and Johnsen (2017), White (2018), and Sharma et al. (2009). Gary Larson's cartoons are in several book collections of his amazing work and literally flood the internet.

Akcali, Christopher K., and David W. Pfennig. "Geographic Variation in Mimetic Precision among Different Species of Coral Snake Mimics." *Journal of Evolutionary Biology* 30, no. 7 (2017): 1420–1428.

Arenas, Lina María, Dominic Walter, and Martin Stevens. "Signal Honesty and Predation Risk among a Closely Related Group of Aposematic Species." *Scientific Reports* 5, no. 1 (2015): 1–12.

Brower, Lincoln Pierson. "Ecological Chemistry." *Scientific American* 220, no. 2 (1969): 22–29.

Cuthill, Innes C., William L. Allen, Kevin Arbuckle, Barbara Caspers, George Chaplin, Mark E. Hauber, Geoffrey E. Hill, et al. "The Biology of Color." *Science* 357, no. 6350 (2017): eaan0221.

Dale, J., C. J. Dey, K. Delhey, B. Kempenaers, and M. Valcu. "The Effects of Life History and Sexual Selection on Male and Female Plumage Colouration." *Nature* http://dx.doi.org/10.1038/nature15509 (2015).

Ehrlich, Paul R., and Peter H. Raven. "Butterflies and Plants: A Study in Coevolution." *Evolution* 18, no. 4 (1964): 586–608.

Gilbert, N., 2015. "Sexual Selection Makes Female Songbirds Drab." *Nature* doi:10.1038/nature.2015.18735\.

Greene, Harry W., and Roy W. McDiarmid. "Coral Snake Mimicry: Does It Occur?" *Science* 213, no. 4513 (1981): 1207–1212.

Heliconius Genome Consortium. Butterfly Genome Reveals Promiscuous Exchange of Mimicry Adaptations among Species. *Nature* 487 (2012): 94–8.

Kronforst, Marcus R., and Riccardo Papa. "The Functional Basis of Wing Patterning in Heliconius Butterflies: The Molecules behind Mimicry." *Genetics* 200, no. 1 (2015): 1–19.

Marshall, Justin, and Sonke Johnsen. "Fluorescence As a Means of Colour Signal Enhancement." *Philosophical Transactions of the Royal Society B: Biological Sciences* 372, no. 1724 (2017): 20160335.

Nijhout, H. Frederik. "Wing Pattern Formation in Lepidoptera: A Model." *Journal of Experimental Zoology* 206, no. 2 (1978): 119–136.

Pieribone, Vincent, David F. Gruber, and Sylvia Nasar. *Aglow in the Dark: The Revolutionary Science of Biofluorescence*. Harvard University Press, 2005.

Prötzel, David, Martin Heß, Mark D. Scherz, Martina Schwager, Anouk van't Padje, and Frank Glaw. "Widespread Bone-Based Fluorescence in Chameleons." *Scientific Reports* 8, no. 1 (2018): 698.

Rabosky, Alison R. Davis, Christian L. Cox, Daniel L. Rabosky, Pascal O. Title, Iris A. Holmes, Anat Feldman, and Jimmy A. McGuire. "Coral

Snakes Predict the Evolution of Mimicry across New World Snakes."
Nature Communications 7 (2016): 11484.

Sharma, Vivek, Matija Crne, Jung Ok Park, and Mohan Srinivasarao.
"Structural Origin of Circularly Polarized Iridescence in Jeweled Beetles."
Science 325, no. 5939 (2009): 449–451.

Stevens, Martin, and Sami Merilaita. "Animal Camouflage: Current Issues
and New Perspectives." *Philosophical Transactions of the Royal Society B:
Biological Sciences* 364, no. 1516 (2008): 423–427.

White, Thomas E. "Cryptic Coloration." (2018). J. Vonk, T. K. Shackelford (eds.),
Encyclopedia of Animal Cognition and Behavior, C, pp1–3.). Heidelberg: Springer.

Stoddard, Mary Caswell, and Richard O. Prum. "How Colorful Are Birds?
Evolution of the Avian Plumage Color Gamut." *Behavioral Ecology* 22, no. 5
(2011): 1042–1052.

Wallace, Alfred Russel. *Mimicry, and Other Protective Resemblances Among
Animals*. London, Read Books Ltd, 2016.

White, Thomas E. "Illuminating the Evolution of Iridescence." *Trends in
Ecology & Evolution* 33, no. 6 (2018): 374–375.

Westerman, Erica L., Nicholas W. VanKuren, Darli Massardo, Ayşe Tenger-
Trolander, Wei Zhang, Ryan I. Hill, Michael Perry, et al. "Aristaless
Controls Butterfly Wing Color Variation Used in Mimicry and Mate
Choice." *Current Biology* 28, no. 21 (2018): 3469–3474.

6. THE COLORS OF HISTORY AND CULTURE

The description of work on the ancient DNA of extinct humans was published
by Taylor and Reimchen (2016). The early development of pigments in human
culture is discussed by Colage and d'Errico (2018) and Tarlach (2018). McCarthy
et al. (2019), Berlin and Kay (1969), and Lindsey and Brown (2009) discuss the
color survey work mentioned in the text. Finally, Victoria Finlay's wonderful book
The Brilliant History of Color in Art can be consulted for a deeper discussion of
color in art. For readers interested in detailed discussions of the technology of color
production we include reference to Hecht (2002) and Falk et al., (1985).

Berlin, Brent, and Paul Kay. *Basic Color Terms*. Berkeley, CA: University of
California Press, 1969.

Colagè, Ivan, and Francesco d'Errico. "Culture: The Driving Force of Human Cognition." *Topics in Cognitive Science* 10: 1–19 (2018).

Falk, David, Dieter Brill & David Stork. *Seeing the light: Optics in nature, photography, color, vision and holography.* Echo Point Books & Media, Brattleboro, VT (2020)

Finlay, Victoria. *The Brilliant History of Color in Art.* Los Angeles: Getty Publications, 2014.

Hecht, Eugene. *Optics*, 4th edition. Pearson Education Limited, Essex (2014).

Lindsey, Delwin T., and Angela M. Brown. "World Color Survey Color Naming Reveals Universal Motifs and Their Within-Language Diversity." *Proceedings of the National Academy of Sciences* 106, no. 47 (2009): 19785–19790.

McCarthy, Arya D., Winston Wu, Aaron Mueller, Bill Watson, and David Yarowsky. "Modeling Color Terminology Across Thousands of Languages." *arXiv preprint arXiv:1910.01531* (2019).

Tarlach, G. "Prehistoric Use of Ochre Can Tell Us About The Evolution of Humans' Cognitive Development." *Discovery Magazine* April, 2018. pp 50–55.

Taylor, John S., and Thomas E. Reimchen. "Opsin Gene Repertoires in Northern Archaic Hominids." *Genome* 59, no. 8 (2016): 541–549.

7. THE COLOR OF HUMANS

This chapter discusses the genetics and sociology of race. Perhaps one of the best authors in this area is Nina Jablonski. We list several of her publications below, including her excellent 2006 book, *Skin: A Natural History.* Four papers (Pavan and Sturm, 2019; Katsara and Nothnagel, 2019; Liu et al., 2009; Hysi et al., 2018) discuss the genetics and genomics of hair and eye color. Albinism and tanning are covered in Jauregui et al. (2018) and Visconti et al. (2018), respectively. A discussion of the concept of race in biology can be found in DeSalle and Tattersall's 2018 book, *Troublesome Science: The Misuse of Genetics and Genomics in Understanding Race.*

DeSalle, Rob, and Ian Tattersall. *Troublesome science: The Misuse of Genetics and Genomics in Understanding Race.* New York: Columbia University Press, 2018.

Hysi, Pirro G., Ana M. Valdes, Fan Liu, Nicholas A. Furlotte, David M. Evans, Veronique Bataille, Alessia Visconti, et al. "Genome-wide Association

Meta-analysis of Individuals of European Ancestry Identifies New Loci Explaining a Substantial Fraction of Hair Color Variation and Heritability." *Nature Genetics* 50, no. 5 (2018): 652–656.

Jablonski, Nina G. "Skin Color." *The International Encyclopedia of Biological Anthropology* (2018): 1–5.

Jablonski, Nina G. "Evolution of Human Skin Color and Vitamin D." In *vitamin D*, pp. 29–44. Cambridge, MA: Academic Press, 2018.

Jablonksi, Nina (2006). Skin: A Natural History. Berkeley, CA: University of California Press.

Jablonski, Nina G., and George Chaplin. "The colours of humanity: the evolution of pigmentation in the human lineage." *Philosophical Transactions of the Royal Society B: Biological Sciences* 372, no. 1724 (2017): 20160349.

Jauregui, Ramon, Laryssa A. Huryn, and Brian P. Brooks. "Comprehensive Review of the Genetics of Albinism." *Journal of Visual Impairment & Blindness* 112, no. 6 (2018): 683–700.

Katsara, Maria-Alexandra, and Michael Nothnagel. "True Colors: A Literature Review on the Spatial Distribution of Eye and Hair Pigmentation." *Forensic Science International: Genetics* 39 (2019): 109–118.

Liu, Fan, Kate van Duijn, Johannes R. Vingerling, Albert Hofman, André G. Uitterlinden, A. Cecile JW Janssens, and Manfred Kayser. "Eye Color and the Prediction of Complex Phenotypes from Genotypes." *Current Biology* 19, no. 5 (2009): R192–R193.

Pavan, William J., and Richard A. Sturm. "The Genetics of Human Skin and Hair Pigmentation." *Annual Review of Genomics and Human Genetics* 20 (2019): 41–72.

Visconti, Alessia, David L. Duffy, Fan Liu, Gu Zhu, Wenting Wu, Yan Chen, Pirro G. Hysi, et al. "Genome-wide Association Study in 176,678 Europeans Reveals Genetic Loci for Tanning Response to Sun Exposure." *Nature Communications* 9, no. 1 (2018): 1–7.

8. THE COLOR OF OUR MINDS

This chapter has two major themes—how colors influence our behaviors and decisions, and how colors help us understand ourselves. The color red is a major driver of our behavior, as Pazda and Elliot (2017), Prokop and Pazda (2016), Gilbert et al. (2016), and Wilms and Oberfeld (2018) describe. The preference of wine drinkers for black and white labels over gaudy colorful wine labels is described by de Mello and Gonçalves (2009). The role of color in understanding perception is discussed in the article "Color Realism and Color Science" published in 2003 by Byrne and Hilbert. Expansion on color realism is addressed in Gatzia (2010) and Hanada (2018). Two books are cited in the discussion of color realism and perception: Hardin (1988) and Chirimuuta (2015). Nigel (2001) emphasizes the hard problem of color perception. Other books cited in this chapter are Locke (1690), Crick (1994), and DeSalle and Tattersall (2012).

Byrne, Alex, and David R. Hilbert. "Color Realism and Color Science." *Behavioral and Brain Sciences* 26, no. 1 (2003): 3–21.

Chirimuuta, Mazviita. *Outside Color: Perceptual Science and the Puzzle of Color in Philosophy.* Cambridge, MA: MIT Press, 2015.

Crick, Francis. *The Astonishing Hypothesis.* New York: Touchstone, 1994.

De Mello, L., and R. Pires Gonçalves. "Message on a Bottle: Colours and Shapes in Wine Labels." Munich Personal RePEc Archive, Paper No. 13122 (2009).

DeSalle, Rob, and Ian Tattersall. *The Brain: Big Bangs, Behaviors, and Beliefs.* New Haven, CT: Yale University Press, 2012.

Gatzia, Dimitria Electra. "Colour Fictionalism." *Rivista DiEstetica* 1, no. 43 (2010): 109–123.

Gilbert, Avery N., Alan J. Fridlund, and Laurie A. Lucchina. "The Color of Emotion: A Metric for Implicit Color Associations." *Food Quality and Preference* 52 (2016): 203–210.

Hanada, Mitsuhiko. "Correspondence Analysis of Color–Emotion Associations." *Color Research & Application* 43, no. 2 (2018): 224–237.

Hardin, Clyde L. *Color for Philosophers: Unweaving the Rainbow.* Cambridge, MA: Hackett Publishing, 1988.

Locke, J. (1690) *An Essay Concerning Human Understanding.* Oxford: Oxford University Press, 1979.

Pazda, A.D., and A. J. Elliot. "Processing the Word Red Can Influence Women's Perceptions of Men's Attractiveness." *Current Psychology* 36 (2017): 316–323.

Prokop, P., and A. D. Pazda. "Women's Red Clothing Can Increase Mate Guarding from Their Male Partner." *Personality and Individual Differences* 98 (2016): 114–117.

Thomas, Nigel J. T. "Color Realism: Toward a Solution to the 'Hard Problem.'" *Consciousness and Cognition* 10 (2001): 140–145.

Wilms, Lisa, and Daniel Oberfeld. "Color and Emotion: Effects of Hue, Saturation, and Brightness." *Psychological Research* 82, no. 5 (2018): 896–914.

EPILOGUE: THE COLOR OF EXISTENCE

Tattersall has written extensively on the origin of consciousnessin our species. Any of his several books can be referred to, but we list here one essay written in 2014.

Tattersall, Ian. "Language as a Critical Factor in the Emergence of Human Cognition." *Humana. Mente Journal of Philosophical Studies* 7, no. 27 (2014): 181–195.

Acknowledgments

The authors would like to acknowledge their respective institutions—The American Museum of Natural History in New York City and Questacon in Canberra, both of which take great care and effort to communicate science to the general public. We would also like to thank Martin Schwabacher, Sasha Nemacek, Cine Ostrow, Lydia Romero, Brett Peterson, and Dina Langis all members of the amazing exhibitions team at the AMNH who created the Nature of Color show for the AMNH. Special thanks goes to Michael Meister whose beautiful design of the Nature of Color show at the AMNH was a great inspiration. Big thanks goes to Lauri Halderman and Melisa Posen who direct the Exhibition Department at the AMNH. We would also like to thank Jessica Case for her expert guidance through the writing and production of the book. The real hard work on the book though, comes from Judy Myers at Pegasus who copy edited and organized our writing and we thank her profusely, as well as Maria Fernandez for her design of the text.

Index